This is a book in the series
STUDIES IN PHYSICS AND CHEMISTRY
Consulting Editors: R. Stevenson and M. A. Whitehead

---

1. *Stevenson* — **MULTIPLET STRUCTURE OF ATOMS AND MOLECULES**
2. *Smart* — **EFFECTIVE FIELD THEORIES OF MAGNETISM**
3. *Fujita* — **INTRODUCTION TO NON-EQUILIBRIUM QUANTUM STATISTICAL MECHANICS**
4. *Pauncz* — **ALTERNANT MOLECULAR ORBITAL METHOD**
5. *Kyrala* — **THEORETICAL PHYSICS: APPLICATIONS OF VECTORS, MATRICES, TENSORS AND QUATERNIONS**
6. *Shadowitz* — **SPECIAL RELATIVITY**
7. *Fraga and Malli* — **MANY-ELECTRON SYSTEMS: PROPERTIES AND INTERACTIONS**
8. *Steele* — **THEORY OF VIBRATIONAL SPECTROSCOPY**
9. *Spencer* — **THE PHYSICS AND CHEMISTRY OF DNA and RNA**

Additional volumes in preparation

STUDIES IN PHYSICS AND CHEMISTRY

Number 9

THE PHYSICS AND CHEMISTRY OF DNA and RNA

by
JOHN H. SPENCER

THE PHYSICS AND CHEMISTRY OF DNA and RNA

JOHN H. SPENCER

Department of Biochemistry
McGill University

QD433
S65
1972

1972

W. B. SAUNDERS COMPANY, Philadelphia, London, Toronto

W. B. Saunders Company:	West Washington Square
Philadelphia, PA 19105

12 Dyott Street
London, WC1A 1DB

833 Oxford Street
Toronto 18, Ontario

The Physics and Chemistry of DNA and RNA					ISBN 0-7216-8513-7

© 1972 by W. B. Saunders Company. Copyright under the International Copyright Union. All rights reserved. This book is protected by copyright. No part of it may be reproduced, stored in a retrieval system, or transmitted in any form or by any means, electronic, mechanical, photocopying, recording, or otherwise, without written permission from the publisher. Made in the United States of America. Press of W. B. Saunders Company. Library of Congress catalog card number 79-183458.

Print No: 9   8   7   6   5   4   3   2   1

PREFACE

Through the years research on the nucleic acids has attracted many physicists and chemists, a number of whom have been responsible for some of the major discoveries concerning these molecules. However, the subject is still considered, incorrectly, a province of the biologist by many scientists. The inclusion of this volume in the physics and chemistry series is an attempt to indicate areas, in this subject, of potential research interest to physicists and chemists.

The volume is not a comprehensive review of the subject, since this would involve a treatise of several volumes. The subject matter covered is devoted mainly to the nucleic acids as macromolecules, with an emphasis on structure and organization. A consideration of the chemistry of the components of the nucleic acids serves as an introduction to the molecules, followed by a discussion of their organization into macromolecules and macromolecular properties. The synthetic polynucleotides are considered separately, including their synthesis, properties and study as model compounds. Finally, the physicochemical methodology used in the study of the nucleic acids and some physicochemical properties are discussed. The biology of the nucleic acids is not included because excellent books are available on the subject.

More references are given than is usual for a volume of this size, and review articles have been referred to wherever possible, to enable the reader to investigate any topic in greater depth. Nevertheless, the selection of references has to be arbitrary and, like the text material, is not comprehensive.

The completion of this volume requires acknowledgment of the excellent secretarial assistance of Mrs. C. J. Hellstrom at McGill University and the staff of the W. B. Saunders Company.

JOHN H. SPENCER

CONTENTS

INTRODUCTION 1
    References 4

CHAPTER 1   STRUCTURAL COMPONENTS OF THE NUCLEIC ACIDS 6
    Sugars 6
    The Nucleic Acid Bases 10
        Base Components of DNA 10
        Base Components of RNA 13
            Pseudouridine 14
            N^6-($\triangle^2$-Isopentenyl) adenosine 15
    References 20

CHAPTER 2   MOLECULAR STRUCTURE OF THE NUCLEIC ACIDS 22
    Occurrence, Composition and Primary Structure 22
        Ribonucleic Acids 22
            Ribosomal RNA 23
            Messenger RNA 24
            Transfer RNA 25
            5 s RNA 27
            Minor Uncharacterized RNA Species 28
            Viral RNA 28

Deoxyribonucleic Acids 29
 Animal and Plant DNA 31
  Nuclear DNA 31
  Satellite DNA 32
  Cytoplasmic DNA 33
 Bacterial DNA 34
 Viral DNA 34
Sequence Studies on the Nucleic Acids 35
 Ribonucleic Acids 35
  Determination of the Sequence of
   $E.\ coli$ 5 s RNA 36
 Deoxyribonucleic Acids 40
Configuration of the Nucleic Acids 42
 Configuration of the Ribonucleic Acids 42
  Configuration of Ribosomal RNA 42
  Configuration of Messenger RNA 47
  Configuration of Transfer RNA 48
  Configuration of 5 s RNA 52
 Configuration of DNA 55
References 62

CHAPTER 3 SYNTHETIC MODEL POLYNUCLEOTIDES .. 69

Synthesis of Model Polynucleotides 69
 Polynucleotide Phosphorylase 69
  Preparation and Assay 70
  Polymerization Reaction 70
  Substrate Specificity 71
  Secondary Reactions 72
 Synthesis by Other Enzymes 73
 Chemical Synthesis of Polynucleotides 73
  Condensation Reactions 74
  Step-wise Synthesis 79
  Synthesis on a Polymeric Support 82
 Multiplication of Preformed Oligodeoxynucleotides
  and Synthesis of RNA of Complementary Sequence 85
 Synthesis of Oligodeoxynucleotides and Joining
  Reactions 85
Studies with Synthetic Model Polynucleotides 88
 Polymer-Polymer Interaction 88
 Monomer and Oligonucleotide-Polymer Interaction 91
 Single-Stranded Polymer Interactions 93
References 95

CHAPTER 4 PHYSICAL METHODOLOGY AND THE MACROMOLECULAR PROPERTIES OF THE NUCLEIC ACIDS 99

Osmotic Pressure 99
Hydrodynamic Properties 101
   Viscosity 101
      Capillary Viscometers 103
      Rotating Cylinder Viscometers 105
      Falling Sphere Viscometers 106
      Viscosity of Macromolecular Solutions 106
   Sedimentation Analysis 108
      Sedimentation Equilibrium Method 109
      Sedimentation Velocity Method 111
      The Archibald Method of Sedimentation Analysis 113
      Density Gradient Centrifugation 114
      Band Sedimentation 115
      Zone-Velocity Sedimentation in Preformed Density Gradients 115
      The Analytical Ultracentrifuge 116
   Relationship Between Sedimentation and Viscosity 117
   Determination of Molecular Weights by Electron Microscopy 120
   Flow Birefringence 122
Spectroscopic Properties 124
   Ultraviolet and Visible Spectra 126
      Thermal Denaturation of Nucleic Acids 128
   Optical Rotatory Dispersion and Circular Dichroism 131
   Infrared Spectra 131
   Magnetic Resonance Spectra 131
      Nuclear Magnetic Resonance Spectra 132
      Electron Spin Resonance 135
References 136

INDEX 139

INTRODUCTION

The genetic material of the cell is DNA. DNA is the template for synthesis of RNA, which in turn is the template for synthesis of protein. The central role of the nucleic acids—DNA and RNA—in the life process has generated a large and continuing scientific effort to determine their structure, function and the control mechanisms by which they operate. Because of the success of this research, the abbreviations DNA and RNA, for deoxyribonucleic acid and ribonucleic acid, have become familiar, everyday terms to all scientists, regardless of discipline.

The history of the study of the nucleic acids is divided into two distinct phases. The first comprises the discovery of and the classic studies on the organic chemistry of the component molecules. The second phase was initiated by the discovery of the biological role of the nucleic acids, which in turn stimulated the current interest in the molecules.

The nucleic acids were discovered more than 100 years ago by Miescher, working in the laboratory of Hoppe-Seyler in 1868 and 1869. Miescher isolated a substance, which he called nuclein, from the nuclei of pus cells and showed that it contained a high proportion of phosphorus. This discovery was of special interest since at that time the only known phosphorus-containing organic compound in tissue was lecithin. Hoppe-Seyler found Miescher's results so surprising that before publishing them in his own journal he repeated the experiments personally. Two years later, in 1871, he published Miescher's original observations[36] together with his own confirmatory results[30] and supplementary experiments by two of his students on casein, egg yoke, yeast and the red cells of birds and reptiles.[34,38]

Miescher returned to Switzerland where he continued his studies on nuclein, and using the sperm heads of the Rhine salmon as a convenient source of material, he isolated a material of high molecular weight. This material had a phosphorus content of 9.59 per cent and gave analytical figures corresponding to what is now known as nucleic acid. Miescher's preparative methods, involving extractions at low temperatures under most exacting conditions, were overlooked by later workers, who made use of material from tissues subjected to heat and then extracted by acid and alkali. These methods yielded a product which, while suitable for the degradative studies and examination of hydrolysis products, bore little resemblance to the native nucleic acid.

Nevertheless, the degradative studies were of fundamental importance in determining the elementary molecules which constitute the nucleic acids. From these studies, two distinct types of nucleic acid were described. The nucleic acid from yeast, on hydrolysis, yielded adenine, guanine, cytosine, uracil, phosphoric acid and a sugar recognized as a pentose by Hammarsten[28] and identified as ribose by Levene and Jacobs.[32] The other, the nucleic acid from thymus gland, yielded adenine, guanine, cytosine, thymine, phosphoric acid and a sugar at first thought to be a hexose, but later identified in Levene's laboratory as deoxyribose.[33] These two nucleic acids came to be called ribonucleic acid and deoxyribonucleic acid, respectively, and since most nucleic acids of animal origin appeared to resemble that from thymus, while those of plant origin were similar to that from yeast, it was assumed that pentose nucleic acids were characteristic of plants and deoxypentose nucleic acids of animal tissues.

This classification was prone to the discovery of exceptions, and its invalidity was proved by the classic histochemical studies of Brachet, using the ribonuclease test,[4-7] the quantitative ultraviolet spectrophotometric examination of tissues by Caspersson[9-16] and the chemical analysis of cells by Davidson and Waymouth.[22,23]

Much of the early information on nucleic acids, since it was derived from degraded material, led to erroneous conclusions regarding size and macromolecular structure and retarded development of structural and functional concepts of nucleic acids. The prime example of this was the tetranucleotide hypothesis, based on the equimolar proportions of the four nucleotides found in hydrolysates of yeast nucleic acid and on an erroneous determination of a molecular weight of 1.3×10^3 consistent with a tetranucleotide structure.[37]

The tetranucleotide hypothesis went unchallenged until the early 1940's, when investigations of molecular weights and compositions of ribonucleic acid preparations indicated that the hypothesis might not be valid. These initial studies led to an emphasis on development of isolation techniques for nucleic acids which did not involve degradation, during the extraction processes.

During this time, the precise analysis of the chemical constituents of cells and tissues with very small amounts of material became possible with the introduction of paper chromatography[19,35,43] and ion exchange chromatography.[20] With the isolation of undegraded nucleic acid preparations, the tetranucleotide "molecular weight" was shown to be erroneous,[25] and the microchemical analyses carried out in Chargaff's laboratory[17,18,35] showed that there are many nucleic acids which differ in composition with regard to molar proportions of bases, depending on the biological source of the material from which they are derived.

The culmination of these chemical studies was a renewed interest in the biological function of the nucleic acids and integration of this function with the structure of the nucleic acids. The concept that the nucleic acids carry the genetic code stems from the discovery by Avery, Macleod and McCarty[2] of bacterial transformation, first described by Griffith in 1928.[27] This work did not immediately receive the widespread recognition that it deserved because ideas concerning the structure of nucleic acids were still prejudiced at that time in favor of the erroneous tetranucleotide structure.

The final stages of the classic investigations which resulted in the clarification of the primary structure of nucleic acids were paralleled by equally dramatic progress toward an understanding of their spatial conformation. Analytical results from Chargaff's laboratory on DNA's of various species revealed a proportional relationship between the molar amounts of the various bases[17] which has since become known as Chargaff's Rule. This rule states that $A = T$ and $G = C$. This information, correlated with the X-ray diffraction studies of Wilkins[42] and Franklin[26] enabled Watson and Crick[40] to postulate a double-stranded helical structure for DNA which provided an explanation of the chemistry of the molecule and its biological role as the carrier of the genetic information.[41] Subsequent work has consistently confirmed and reinforced their postulated model, which has required only minor modification and refinement.

Thus, the study of the nucleic acids developed into its second phase, made possible only by the development and refinement of a number of physicochemical techniques for establishing the gross size and shape of macromolecules in solution. These include light scattering, ultracentrifugation, viscometry, birefringence analysis and electron microscopy. To probe the detailed fine structure of the nucleic acids, such techniques as X-ray diffraction, infrared spectroscopy, optical rotatory dispersion, circular dichroism, nuclear magnetic resonance spectroscopy and electron spin resonance spectroscopy have been utilized. These studies have been complemented by the equally important applications of radioactive tracer studies of the biosynthesis of the molecules, enzymology for synthesis of model polynucleotides *in vitro*, high resolution electron microscopy coupled with biosynthetic and degradative experiments, and chemical modification of the molecules.

In 1965, the first complete nucleotide sequence of a small ribonucleic acid molecule, alanine transfer RNA, was determined[29]. Sequences for a number of other transfer RNA molecules have since been elucidated, as well as for the slightly larger 5 s RNA.[8] Sequence data on transfer RNA's has led to studies of secondary and tertiary structure. Various RNA sequencing techniques are currently being applied to viral RNA molecules[3,21,31,39] and ribosomal RNA.[24]

The most recent major advance in nucleic acid chemistry has been the synthesis in Khorana's laboratory of the DNA sequence corresponding to the gene for alanine transfer RNA.[1]

4 / INTRODUCTION

The sequence and synthetic work provides an excellent basis for current studies on structure-function relationships at the molecular level of organization. Perhaps this will herald the beginning of a third phase in nucleic acid research.

REFERENCES

1. Agarwal, K. L., Büchi, H., Caruthers, M. H., Gupta, N., Khorana, H. G., Kleppe, K., Kumar, A., Ohtsuka, E., RajBhandary, U. L., van de Sande, J. H., Sgaramella, V., Weber, H., and Yamada, T., Nature *227*, 27 (1970).
2. Avery, O. T., MacLeod, C. M., and McCarty, M., J. Exptl. Med. *79*, 137 (1944).
3. Billeter, M. A., Dahlberg, J. E., Goodman, H. M., Hindley, J., and Weissman, C., Cold Spring Harb. Symp. Quant. Biol. *34*, 635 (1969).
4. Brachet, J., Arch. Biol. (Liège) *44*, 519 (1933).
5. Brachet, J., Arch. Biol. (Liège) *48*, 529 (1937).
6. Brachet, J., Arch. Biol. (Liège) *51*, 151 (1940), *51*, 167 (1940).
7. Brachet, J., C. R. Soc. Biol. (Paris) *133*, 88 (1940), *133*, 90 (1940).
8. Brownlee, G. G., Sanger, F., and Barrell, B. G., J. Mol. Biol. *34*, 379 (1968).
9. Caspersson, T., Skand. Arch. Physiol. *74*, supplement 8 (1936).
10. Caspersson, T., J. Roy. Microscop. Soc. *60*, 8 (1940).
11. Caspersson, T., Naturwissenschaften *29*, 33 (1941).
12. Caspersson, T., Nyström, C., and Santesson, L., Naturwissenschaften *29*, 29 (1941).
13. Caspersson, T., Nyström, C., and Santesson, L., Acta Radiol. (Stockholm) supplement 46 (1942).
14. Caspersson, T., and Schultz, J., Nature *143*, 602 (1939).
15. Caspersson, T., and Schultz, J., Proc. Nat. Acad. Sci. (Wash.) *26*, 507 (1940).
16. Caspersson, T., and Thorell, B., Chromosoma *2*, 132 (1941).
17. Chargaff, E., Experientia *6*, 201 (1950).
18. Chargaff, E., J. Cell. Comp. Physiol. *38*, supplement 1, 41 (1951).
19. Chargaff, E., in *The Nucleic Acids* (E. Chargaff and J. N. Davidson, Eds.), Academic Press, New York, Vol. 1, p. 307 (1955).
20. Cohn, W. E., in *The Nucleic Acids* (E. Chargaff and J. N. Davidson, Eds.), Academic Press, New York, Vol. 1, p. 211 (1955).
21. Cory, S., Spahr, P. F., and Adams, J. M., Cold Spring Harb. Symp. Quant. Biol. *35*, 1 (1970).
22. Davidson, J. N., and Waymouth, C., Nature *152*, 47 (1943).
23. Davidson, J. N., and Waymouth, C., Biochem. J. *38*, 39 (1944), *38*, 375 (1944), *38*, 379 (1944).
24. Fellner, P., Ehresmann, C., and Ebel, J. P., Cold Spring Harb. Symp. Quant. Biol. *35*, 29 (1970).
25. Fletcher, W. E., Thesis, University of London (1948).
26. Franklin, R. E., and Gosling, R. G., Nature *171*, 740 (1953).
27. Griffith, F., J. Hyg., *27*, 113 (1928).
28. Hammarsten, O., Z. Physiol. Chem. *19*, 19 (1894).
29. Holley, R. W., Apgar, J., Everett, G. A., Madison, J. T., Marquisee, M., Merrill, S. H., Penswick, J. R., and Zamir, A., Science *147*, 1462 (1965).
30. Hoppe-Seyler, F., Hoppe-Seyler's Med. Chem. Unters. 486 (1871).
31. Jeppesen, P. G. N., Nichols, J. L., Sanger, F., and Barrell, B. G., Cold Spring Harb. Symp. Quant. Biol. *35*, 13 (1970).
32. Levene, P. A., and Jacobs, W. A., Ber. dtsch. chem. Ges. *42*, 2102 (1909), *42*, 2469 (1909), *42*, 2474 (1909), *42*, 2703 (1909).
33. Levene, P. A., Mikeska, L. A., and Mori, T., J. Biol. Chem. *85*, 785 (1930).
34. Lübavin, N., Hoppe-Seyler's Med. Chem. Unters. 463 (1871).
35. Magasanik, B., in *The Nucleic Acids* (E. Chargaff and J. N. Davidson, Eds.), Academic Press, New York, Vol. 1, p. 373 (1955).
36. Miescher, F., Hoppe-Seyler's Med. Chem. Unters. 441 and 502 (1871).

37. Myrbäck, K., and Jorpes, E., Hoppe-Seyler's Z. *237*, 159 (1935).
38. Plósz, P., Hoppe-Seyler's Med. Chem. Unters. 461 (1871).
39. Steitz, J. A., Cold Spring Harb. Symp. Quant. Biol. *34*, 621 (1969).
40. Watson, J. D., and Crick, F. H. C., Nature *171*, 737 (1953).
41. Watson, J. D., and Crick, F. H. C., Nature *171*, 964 (1953).
42. Wilkins, M. H. F., Stokes, A. R., and Wilson, H. R., Nature *171*, 738 (1953).
43. Wyatt, G. R., in *The Nucleic Acids* (E. Chargaff and J. N. Davidson, Eds.), Academic Press, New York, Vol. 1, p. 243 (1955).

Chapter 1 STRUCTURAL COMPONENTS OF THE NUCLEIC ACIDS

The basic primary structure of the nucleic acids consists of pentose sugar moieties linked together by 3′, 5′ phosphodiester bonds, forming a sugar-phosphate backbone with the nucleic acid bases attached to the C-1′ positions of the sugar moieties through N-glycoside bonds (Figure 1–1). The evidence supporting this structure has been fully described in a number of reviews and textbooks.[7,23,40] In recent years investigations of the structural components of the nucleic acids have centred on the isolation and determination of the structure of minor components discovered in various nucleic acids.

Sugars

Two 5-carbon sugars have so far been shown to be present in nucleic acids—D-ribose in RNA and 2-deoxy-D-ribose in DNA. Both pentoses exist in the furanose form when incorporated into nucleic acids and the glycosidic linkage to the base has the β-configuration.

Early work suggested the possibility of 2′, 5′ phosphodiester links between ribose moieties of RNA, based on the isolation of nucleoside 2′-monophosphates from alkaline hydrolysates of RNA. These compounds were not present in spleen phosphodiesterase digests, and it has since been shown that the 2′ phosphate compounds result from a cyclization reaction occurring during alkaline hydrolysis, as shown in Figure 1–2.[5,6,30] The possibility of 2′ phosphate esterification on ribose also led to numerous speculations regarding the possibility of branching in RNA, but more recent studies have almost eliminated this possibility.[24] In DNA the absence of the 2′ hydroxyl group in deoxyribose did not allow speculation with regard to this type of branching in the molecule.

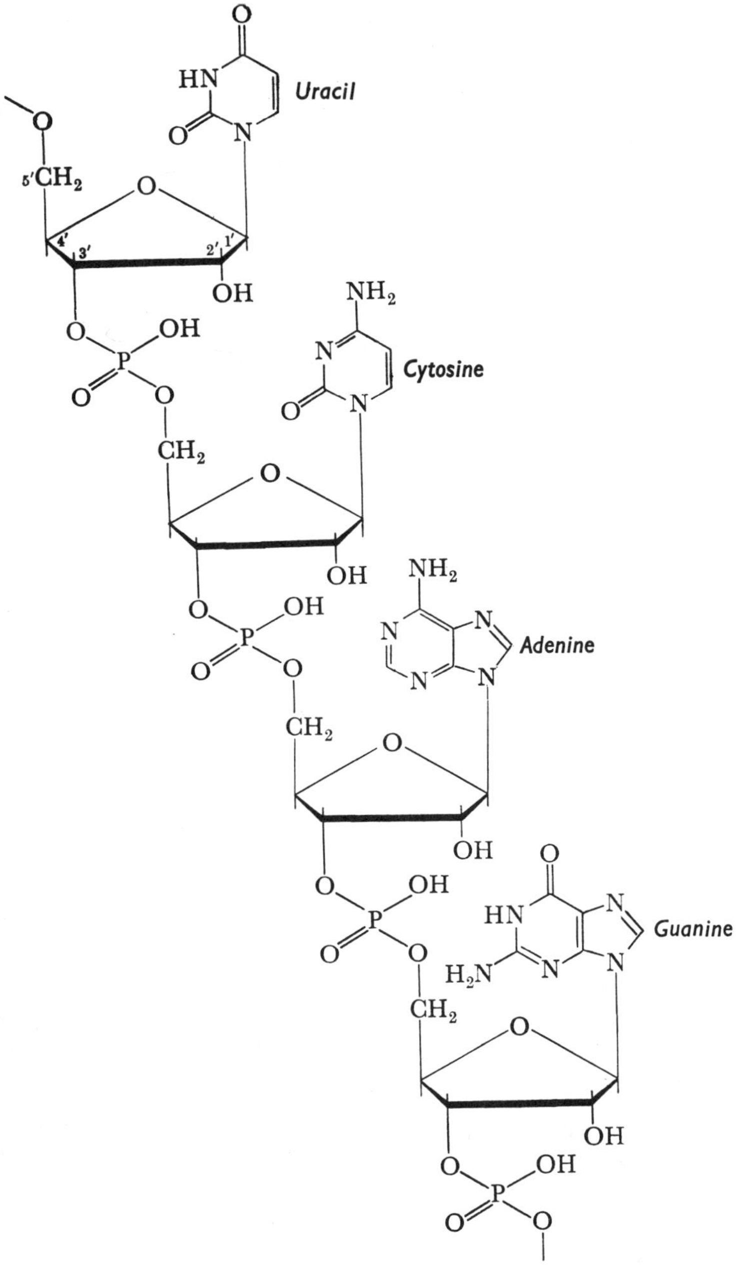

STRUCTURAL COMPONENTS OF THE NUCLEIC ACIDS / 7

Figure 1-1.

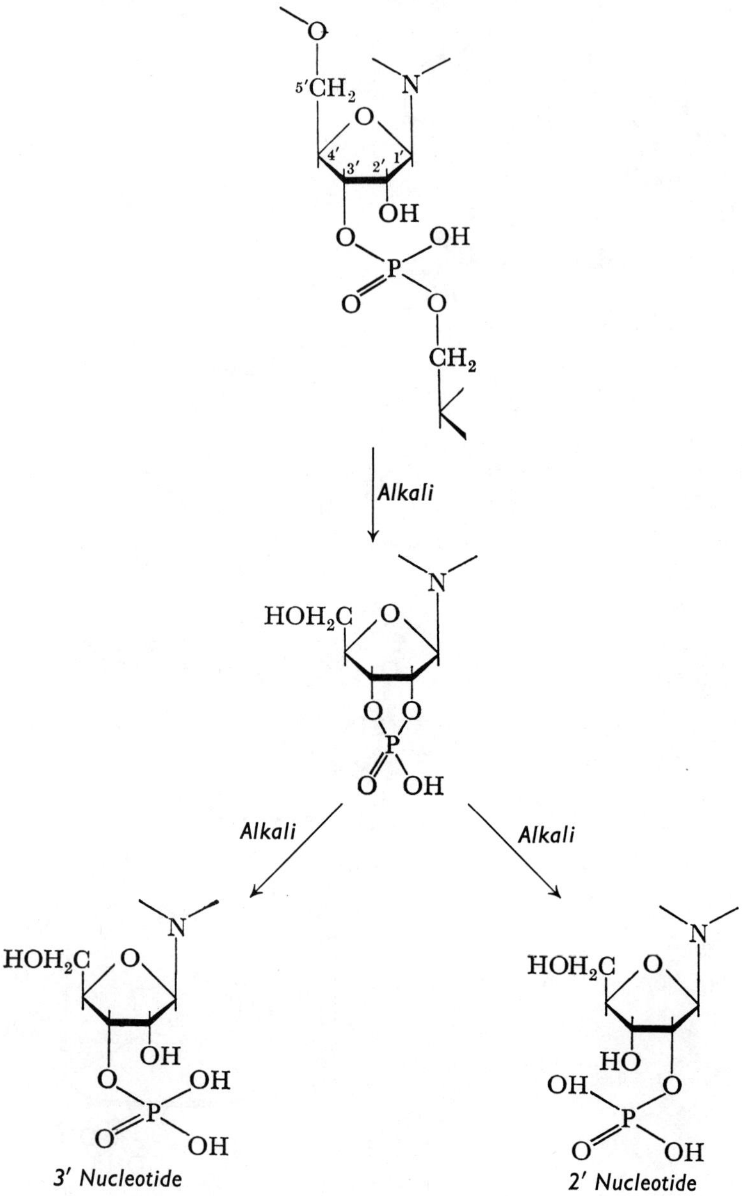

8 / STRUCTURAL COMPONENTS OF THE NUCLEIC ACIDS

Figure 1-2.

STRUCTURAL COMPONENTS OF THE NUCLEIC ACIDS / 9

Ribose also occurs to a small extent as the 2'-O-methyl derivative (I)

I

2'-O-Methyl-I-β-D-ribofuranosyladenine

in some types of RNA.[17,38] 2'-O-Methylribose provides alkali stability at its site of incorporation into RNA molecules since the 2' hydroxyl group cannot participate in a phosphate cyclization reaction. This property has been a major factor in its discovery, isolation and characterization. Most of the 2'-O-methyl-D-ribose found in RNA has been located in transfer RNA, a low molecular weight RNA (2.5×10^4 daltons, 72 to 82 nucleotides), which functions biologically as a carrier of amino acids to the site of protein synthesis in the cell and in ribosomal RNA, the highest molecular weight RNA (1.2×10^6 daltons) present in the cell.[18] Ribosomal RNA is present in the cell associated with protein in the form of discrete particles called ribosomes, which provide a site for the interaction of messenger RNA (the genetic translation material) and transfer RNA by means of a stereochemical relationship for translation of the genetic message of messenger RNA into protein. No report of the occurrence of 2'-O-methylribose in messenger RNA has been made so far. The biological significance of 2'-O-methylribose derivatives in RNA is unknown.

Glucose residues occur in some specific bacteriophage DNA's as part of the side chains of modified cytosine residues. The structure of these compounds is described in the following section on nucleic acid bases.

The occurrence in sponges of free nucleosides in which the sugar moiety is arabinose (IIa and IIb) was first described in the early 1950's.[2,3] Their natural occurrence is so far limited to one species of sponge. These compounds have recently come under study because of their pharmacological properties, particularly with regard to inhibition of leukemias and antiviral activity. Arabinose-containing nucleotides are incorporated into nucleic acid molecules *in vitro*,[10,36] but evidence that they are components of naturally occurring nucleic acids is lacking. The arabinosides have been the subject of a review by Cohen.[11] Since they do not appear to be an important

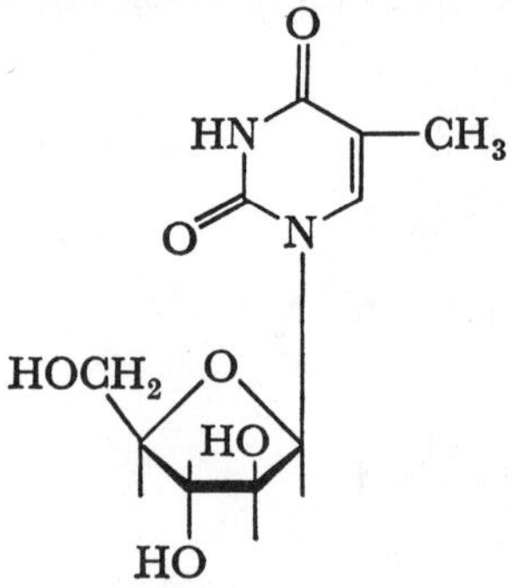

IIa
Spongothymidine
1-β-D-Arabinofuranosylthymine

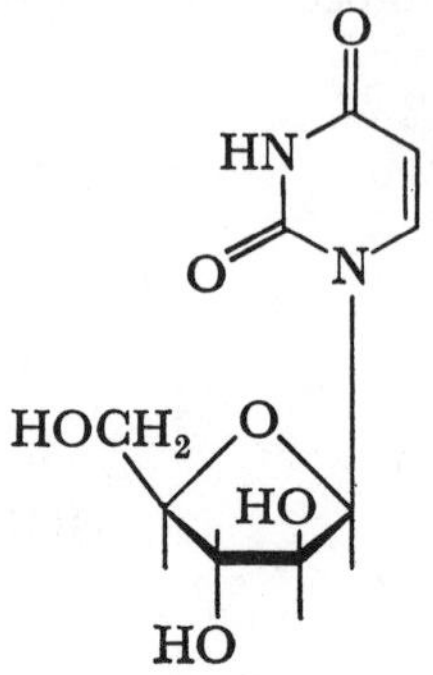

IIb
Spongouridine
1-β-D-Arabinofuranosyluracil

factor in nucleic acids at the macromolecular level, they will not be discussed further in this monograph.

THE NUCLEIC ACID BASES

Base Components of DNA

The four major base components of DNA are the purines adenine and guanine, and the pyrimidines cytosine and thymine (5-methyluracil). In double-stranded DNA molar proportions of the bases occur so that purines = pyrimidines, and that A = T and G = C. A number of minor components have also been shown to occur in certain DNA molecules. All the minor components are derivatives of one of the four major base components and do not alter the A = T, G = C rule. The most widely occurring of the minor bases is 5-methylcytosine (III), which comprises up to 6 per cent of the

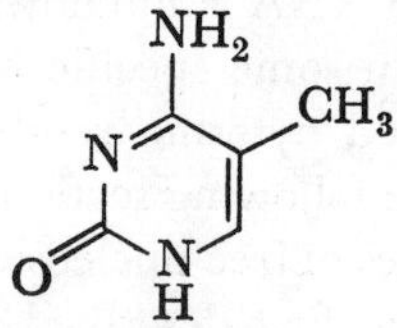

III
5-Methylcytosine

total base composition of plant DNA's[39,43,47] and up to 1.5 per cent of the base composition of certain mammalian DNA's.[39,45] It is also present in the DNA of some bacteria[15,44] and bacteriophages (bacterial viruses).[26,31,33] Other modified bases which are also derivatives of cytosine include 5-hydroxymethylcytosine (IV),[48] monoglucosylated 5-hydroxymethylcytosine

THE NUCLEIC ACID BASES / 11

IV
5-Hydroxymethylcytosine

and diglucosylated 5-hydroxymethylcytosine (V).[22,37,46] These components

V
5-Hydroxy-1-α-(6-O-β-D-glucopyranosyl-D-glucose)methylcytosine

have been shown to be present in the DNA of the T-even bacteriophages (T_2, T_4 and T_6); in these phages the cytosine component of the DNA is substituted entirely by any one or a mixture of 5-hydroxymethylcytosine and the monoglucosylated and diglucosylated 5-hydroxymethylcytosine derivatives.[28] The configuration of the glycoside linkage between the glucose residue and the hydroxymethyl group of the nucleic acid base has been shown to occur in both the α and β configurations, and both configurations occur in T-even bacteriophages.[27] The variations are shown in Table 1-1.

Table 1-1. 5-hydroxymethylcytosine Nucleotides Present in T-even Bacteriophage DNA

Phage	Non-glucosylated	Mono-glucosylated	Glycoside Linkage	Diglucosylated	Glycoside Linkage
T_2	24	70	α	6	α, β
T_4	0	100	70% α, 30% β	0	—
T_6	25	3	α	72	α, β

Another minor base reported as a constituent of DNA is 5-hydroxymethyluracil (VI). Thymine itself is a derivative of uracil (5-methyluracil) and its 5-hydroxy derivative has been shown to replace thymine in bacteriophages SP8[25] and φe[35] of *Bacillus subtilis*. There has been no report of glucose attached to 5-hydroxymethyluracil in DNA. Deoxyuridine has been

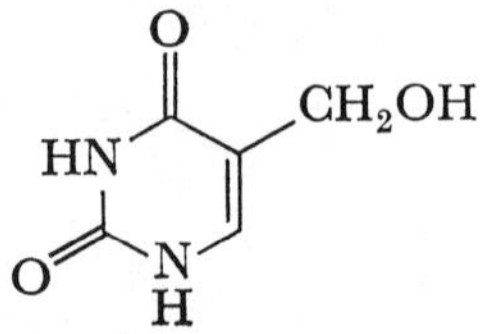

$$\text{VI}$$

5-Hydroxymethyluracil

shown to replace thymine in the transducing phage PBS2.[41] The purine derivative, N^6-methyladenine (VII), is a component of some bacterial[16,44]

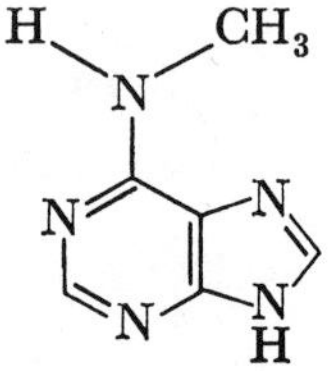

$$\text{VII}$$

6-Methyladenine

and bacteriophage DNA's[16,31] and has been reported present in bovine and human sperm DNA[42] together with N^2-methylated derivatives of guanine. However, it has not been found in other mammalian or plant DNA's.[16,45]

Table I-2. Minor Nucleoside Components Found in Transfer Ribonucleic Acids

NUCLEOSIDE	YEAST TRANSFER RNA	NUMBER OF RESIDUES PER MOLECULE
N^6-Methyladenosine	Tyrosine, phenylalanine	1
N^6-(Δ^2-Isopentenyl)-adenosine	Serine, tyrosine	1
Inosine	Alanine, serine	1
1-Methylinosine	Alanine	1
5,6-Dihydrouridine	Alanine, serine, tyrosine, phenylalanine	2, 3, 6, 2
Pseudouridine	Alanine, serine, tyrosine, phenylalanine	2, 3, 3, 2
2'-O-Methyluridine	Serine	1
Ribosylthymine	Alanine, serine, tyrosine, phenylalanine	1
1-Methylguanosine	Alanine, tyrosine, phenylalanine	1
N^7-Methylguanosine	Phenylalanine	1
N^2-Dimethylguanosine	Alanine, serine, tyrosine, phenylalanine	1
2'-O-Methylguanosine	Serine, tyrosine, phenylalanine	1
5-Methylcytidine	Serine, tyrosine, phenylalanine	1, 1, 2
2'-O-Methylcytidine	Phenylalanine	1
N^4-Acetylcytidine	Serine	1

Data were obtained from the sequences for yeast transfer RNA's; alanine,[21] serine,[49] tyrosine[29] and phenylalanine.[32]

Base Components of RNA

The four major base components of RNA are adenine, guanine, cytosine and uracil. In the past few years a relatively large number of minor base components have been reported to be present in RNA, particularly in the low molecular weight ribonucleic acid transfer RNA. All the minor base components described so far are derivatives of the four major base components, and some of those found in transfer RNA are listed in Table 1–2. However, the most commonly occurring minor component is not a modified base but is a nucleoside, pseudouridine, in which the base uracil is linked from the C-5 position to the C-1′ position of the ribose by a C—C bond (VIII). One of the more recently described minor components is N^6-

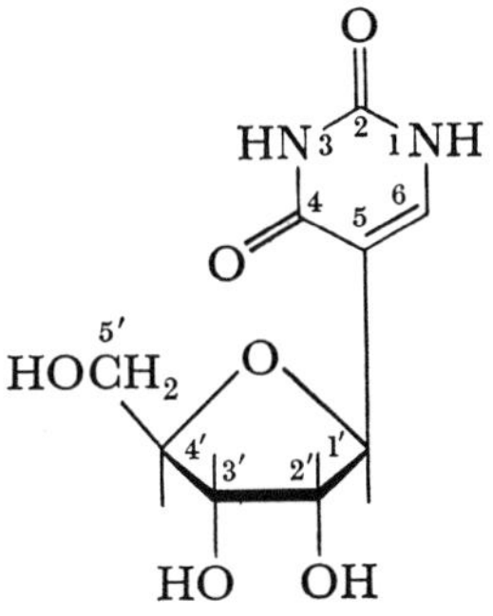

VIII

5-β-D-Ribofuranosyluracil
(pseudouridine)

(Δ^2-isopentenyl)adenosine (IX). The type of studies necessary for the

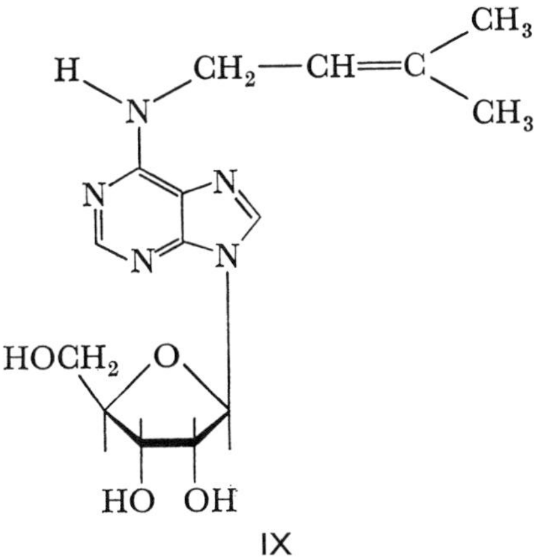

IX

6-N-(3-Methyl-but-2-enylamino)-9-β-D-ribofuranosylpurine
or
N^6-(Δ^2-Isopentenyl)adenosine

Pseudouridine

Pseudouridine was first reported by Davis and Allen in 1957,[14] and its structure was first described by Cohn in 1959.[12] The compound first appeared as an unknown peak in a column chromatographic separation of an alkaline digest of supernatant fraction RNA (transfer RNA). It has since been shown that one to three residues of pseudouridine occur in transfer RNA molecules, accounting for approximately 4 per cent of the total nucleotide composition.[21,29,32,49] It has not been reported to be a constitutent of other types of RNA.

The initial evidence that uracil was a component of the unknown compound came from spectral shifts with changes in pH in the ultraviolet spectrum, which are characteristic of uracil derivatives. The spectral properties were particularly analogous to 5-hydroxymethyluracil. The ion exchange behavior was similar to that of uridine and periodate oxidation, and borate complex formation proceeded to the same degree as with uridine. Snake venom phosphatase digests of RNA released the nucleoside to which two methyl groups could be added by treatment with diazomethane. The C:N:P ratio of 9:2:1 indicated that no amino group was present. However, the compound was not susceptible to acid hydrolysis under conditions which normally release nucleic acid bases (e.g., perchloric acid or formic acid), and ribose could not be released by prolonged acid hydrolysis, hydrogenation or bromination followed by alkali and acid.

Periodate oxidation and nuclear magnetic resonance spectra gave Cohn the first true indication of the structure.[12,13] The nucleoside reduced one mole of periodate, and the product treated with excess sodium borohydride gave a second compound which again reduced one mole of periodate. When further reduced with sodium borohydride, this was found to be identical chromatographically and spectrophotometrically with 5-hydroxymethyluracil (Figure 1-3). Comparison of the nuclear magnetic resonance spectra of a variety of natural and synthetic analogues of uracil, such as uridine, thymine, 6-methyluracil, 5-hydroxymethyluracil and 5-hydroxyuridine, indicated that the unknown compound had a C-6 proton, no C-5 proton, a change in the character of the C-6 proton from that of uridine to that of 5-substituted uracils, a change in the character of the C-1' proton from that found in uridine (N-C-1') toward that expected of a carbon-linked CH group, and finally a CH_2 group at C-5' similar to that of uridine. From this evidence Cohn postulated the nucleoside to be 5-ribosyluracil.

The proof that ribose was the sugar and in the furanosyl form was indirect. The presence of 2' and 3' pseudouridylates in alkaline hydrolysates, the appearance of only the 3' nucleotide following ribonuclease hydrolysis and the presence of a 5' pseudouridylate in snake venom phosphodiesterase hydrolysates and of pseudouridine in total snake venom hydrolysates all indicated the presence of the usual 2', 3' and 5' hydroxyl

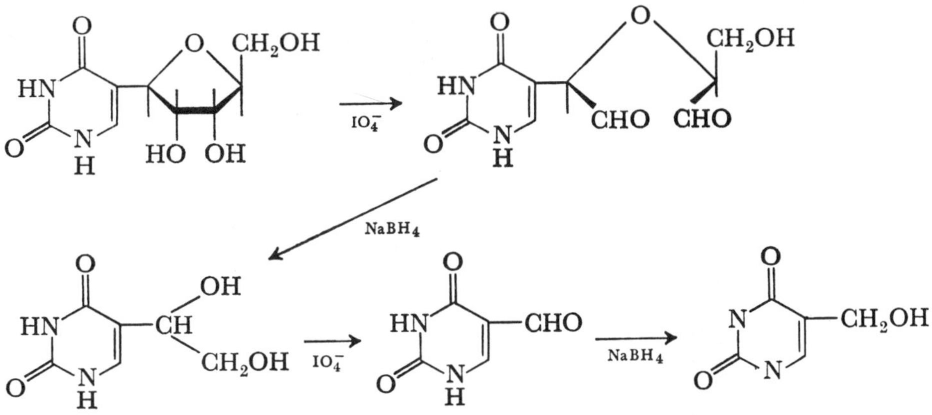

Figure 1-3.

groups and a 3′–5′ phosphodiester linkage. The 40 to 60 ratio of 2′ and 3′ nucleotides usually found in alkaline hydrolysates of RNA was reversed, indicating some new influence upon the cyclic phosphate intermediate. Other indirect evidence concerned the borate and periodate reactions, which were similar both in degree and rate to those for uridine, and consistent with the *cis*-glycol configuration of a furanose structure.

Conclusive evidence of the overall structure was obtained from the nuclear magnetic resonance behavior of the molecule, particularly by the presence of a proton at the C-6 position and the absence of one at C-5 in the pyrimidine portion of the molecule and by the identity of the spectra in the C-2′, C-3′, C-4′ and C-5′ regions in the ribose moiety. Attachment of the C-1′ of ribose to carbon rather than nitrogen was shown by the large positive shift in the C-1′H position relative to uridine. The spectra also showed the absence of a methyl group.

The presence of a

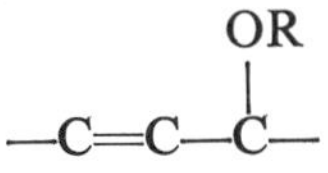

$$-C=C-\underset{\underset{OR}{|}}{C}-$$

grouping in the proposed structure explained the uptake of one mole of hydrogen (by rupture of the furanose bridge) without saturation of the pyrimidine double bond, which occurs with most pyrimidines and results in loss of the characteristic ultraviolet absorption.

Cohn's use of nuclear magnetic resonance spectroscopy for the proof of structure of this compound was one of the early uses of this technique in nucleic acid chemistry and served to demonstrate its usefulness and superiority over previously employed chemical methods of analysis.

N^6-(Δ^2-Isopentenyl)adenosine

N^6-(Δ^2-Isopentenyl)adenosine (IPA) has been described independently by groups from the laboratories of Hall[19] and Zachau[4] as a component of transfer RNA's. Since the original discovery of the molecule, it has been

16 / STRUCTURAL COMPONENTS OF THE NUCLEIC ACIDS

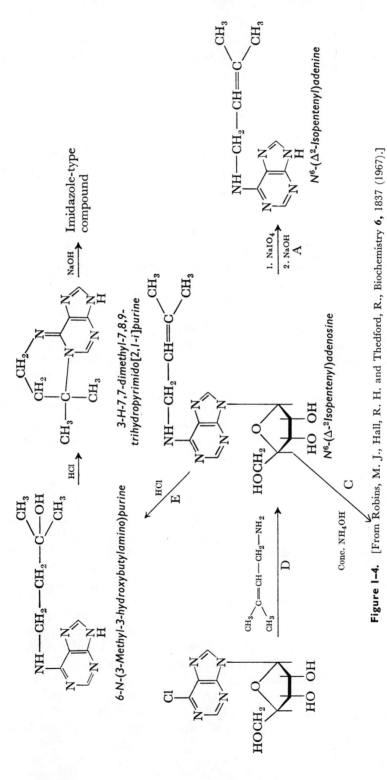

Figure 1-4. [From Robins, M. J., Hall, R. H. and Thedford, R., Biochemistry **6**, 1837 (1967).]

THE NUCLEIC ACID BASES / 17

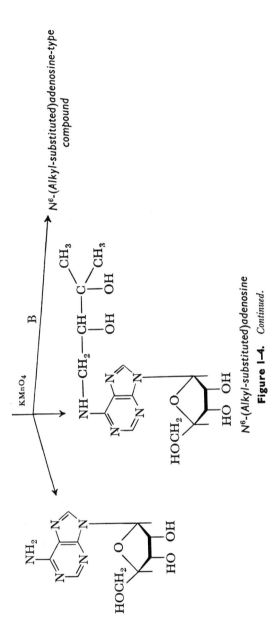

Figure 1-4. *Continued.*

located in a specific sequence of two yeast transfer RNA's, seryl transfer RNA's I and II, by Zachau, Dütting and Feldman.[49]

The compound has been characterized and chemically synthesized by Robins, Hall and Thedford.[34] They prepared IPA from yeast transfer RNA by snake venom and alkaline phosphatase digestion, separating the IPA from the hydrolysate by partition chromatography on celite eluted with ethyl acetate-saturated water. The isolated nucleoside was susceptible to sodium periodate oxidation and alkaline digestion (reaction A, Figure 1–4), releasing the free base, which indicated that the ribose was in the furanose form and linked to the base by an N-glycoside bond. The free base was characterized by co-chromatography with a synthetic sample of [6-N-(3-methyl-but-2-enylamino)-purine] in two solvent systems. The isolated base material had identical ultraviolet absorption spectra with those of the authentic sample of the synthetic base. The synthetic sample of the base [6-N-(3-methyl-but-2-enylamino)-purine] was prepared by reaction of 6-chloropurine and 3-methyl-but-2-enylamine-1. The synthetic sample had a melting point of 212° to 214°C and an elemental analysis corresponding to $C_{10}H_{13}N_5$ and its structure was checked by nuclear magnetic resonance spectroscopy.

IPA was synthesized by two methods—first from 6-chloro-9-β-D-ribofuranosylpurine and 3-methyl-but-2-enylanine (reaction D, Figure 4), yielding a compound with a melting point of 145° to 147°C and an elemental analysis corresponding to $C_{15}H_{21}N_5O_4$; second from adenosine and γ,γ-dimethylallyl bromide in a N,N-dimethylacetamide solvent system. The product of this reaction was purified by chromatography then crystallized, and had a melting point of 148°C.

Oxidation of IPA with potassium permanganate gave three reaction products—adenosine, a second compound identified as N^6-(alkyl-substituted) adenosine and a third compound of similar structure which was not characterized (reaction B, Figure 4). The alkyl-substituted adenosine had a melting point of 120° to 121°C with an elemental structure of $C_{15}H_{23}N_5O_6$. This compound had a similar absorption spectrum to that of IPA. Mass spectra of the compound indicated that two hydroxyl groups had been inserted at the allylic double bond position of the isopentenyl side chain.

IPA reacts with concentrated ammonium hydroxide to produce adenosine (reaction C, Figure 1–4). No other ultraviolet-absorbing products were detected. Hydrolysis of IPA with HCl (reaction E, Figure 4) resulted in the production of [6-N-(3-methyl-3-hydroxybutylamino)purine], which was in turn hydrolyzed to 3-H-7,7-dimethyl-7,8,9-trihydropyrimido[2,1-i]-purine. Both compounds were isolated, crystallized and characterized by elemental composition, ultraviolet absorption spectroscopy and nuclear magnetic resonance spectroscopy. When treated with sodium hydroxide, the second product gave an imidazole type of compound.

Reaction of IPA with iodine resulted in production of a primary product which in turn degraded to other products. The nuclear magnetic resonance spectrum of the primary product indicated that it consisted of two compounds, based on the fact that two single peaks appeared at $\delta 1.68$ and 1.84 and accounted together for six protons. A singlet at $\delta 8.57$ was assumed to represent the C-2 and C-8 protons. Ultraviolet absorption spectra of the iodinated product were similar to those of N^1,N^6-dimethyladenosine, which suggests that the product consists of two closely related compounds with structures of the type represented below (X).

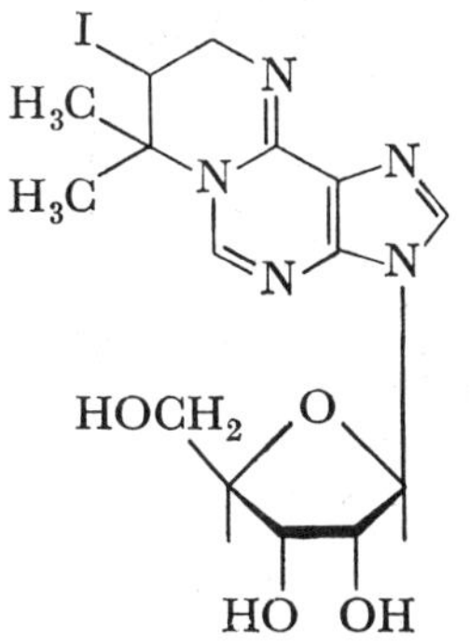

X

Iodinated product of IPA

Thus, from nuclear magnetic resonance spectral considerations and the chemical characterization by means of periodate and sodium hydroxide degradation, $KMnO_4$ oxidation, iodination and alkali and acid degradation, Robins, Hall and Thedford concluded that the structure of IPA was N^6-(Δ^2-isopentenyl)adenosine.

This structural confirmation is another example of the application of modern chemical techniques to biological materials. Many different methylated nucleosides in transfer RNA have been reported (see Table 1–2), and some thiobases such as 2-thiocytosine and 5-methylaminomethyl-2-thiouracil,[9] 2-methylthio-N^6-(Δ^2-isopentenyl)adenosine,[20] 2-thio-5 (or 6)-uridine acetic acid methyl ester,[1] and the cytokinin 6-N-(3-methylbut-2-enylamino)-2-methylthio-9-β-D-ribofuranosylpurine,[8] together with compounds such as 2'-O-ribosyladenosine, N-6-aminoacyladenosine and N-4-acetylcytidine have also been reported. In phenylalanine transfer RNA, the sequence of which has been determined,[32] an unknown nucleoside, which has been designated Y and which is most probably another adenosine derivative, remains to be identified. As the sequences of more different transfer RNA's are determined, reports of more unusual minor components appear. The chances that more unknown bases occur in RNA, particularly transfer RNA, is high. As they are discovered, each of these new bases will require careful characterization in order that the exact alteration in structure from the parent base may be determined.

REFERENCES

1. Baczynskyi, L., Biemann, K., and Hall, R. H., Science *159*, 1481 (1968).
2. Bergmann, W., and Feeney, R. J., J. Amer. Chem. Soc. *72*, 2809 (1950).
3. Bergmann, W., and Feeney, R. J., J. Org. Chem. *16*, 981 (1951).
4. Biemann, K., Tsunakawa, S., Sonnenbichler, J., Feldman, H., Dütting, D., and Zachau, H. G., Angew. Chem. *78*, 600 (1966).
5. Brown, D. M., Magrath, D. I., and Todd, A. R., J. Chem. Soc. 2708 (1952).
6. Brown, D. M., and Todd, A. R., J. Chem. Soc. 2040 (1953).
7. Brown, D. M., and Todd, A. R., in *The Nucleic Acids* (E. Chargaff and J. N. Davidson, Eds.), Academic Press, New York, Vol. 1, p. 409 (1955).
8. Burrows, W. J., Armstrong, D. J., Skoog, F., Hecht, S. M., Boyle, J. T. A., Leonard, N. J., and Occolowitz, J., Science *161*, 691 (1968).
9. Carbon, J., David, H., and Studier, M. H., Science *161*, 1146 (1968).
10. Chu, M. Y., and Fischer, G. A., Biochem. Pharmacol. *14*, 333 (1965).
11. Cohen, S. S., Prog. Nucleic Acid Res. Mol. Biol. *5*, 1 (1966).
12. Cohn, W. E., Biochim. Biophys. Acta *32*, 569 (1959).
13. Cohn, W. E., J. Biol. Chem. *235*, 1488 (1960).
14. Davis, F. F., and Allen, F. W., J. Biol. Chem. *227*, 907 (1957).
15. Doskočil, J., and Šormová, A., Biochim. Biophys. Acta *95*, 513 (1965).
16. Dunn, D. B., and Smith, J. D., Biochem. J. *68*, 627 (1958).
17. Hall, R. H., Biochim. Biophys. Acta *68*, 278 (1963).
18. Hall, R. H., Biochemistry *3*, 876 (1964).
19. Hall, R. H., Robins, M. J., Stasiuk, L., and Thedford, R., J. Amer. Chem. Soc. *88*, 2614 (1966).
20. Harada, F., Gros, H. J., Kimura, F., Chang, S. H., Nishimura, S., and RajBhandary, U. L., Biochem. Biophys. Res. Commun. *33*, 299 (1968).
21. Holley, R. W., Apgar, J., Everett, G. A., Madison, J. T., Marquisee, M., Merrill, S. H., Penswick, J. R., and Zamir, A., Science *147*, 1462 (1965).
22. Jesaitis, M. A., Microbiol. Genet. Bull. *10*, 16 (1954).
23. Jordan, D. O., *The Chemistry of Nucleic Acids*, Butterworth and Co., Ltd., London (1960).
24. Jordan, D. O., *The Chemistry of Nucleic Acids*, Butterworth and Co., Ltd., London, p. 140 (1960) (review).
25. Kallen, R. G., Simon, M., and Marmur, J., J. Mol. Biol. *5*, 248 (1962).
26. Ledinko, N., J. Mol. Biol. *9*, 834 (1964).
27. Lehman, I. R., and Pratt, E. A., J. Biol. Chem. *235*, 3254 (1960).
28. Lichtenstein, J., and Cohen, S. S., J. Biol. Chem. *235*, 1134 (1960).
29. Madison, J. T., Everett, G. A., and Kung, H., Science *153*, 531 (1966).
30. Markham, R., and Smith, J. D., Biochem. J. *52*, 552 (1952).
31. Nikolskaya, I. I., Tkatcheva, Z. G., Vanyushin, B. F., and Tikchonenko, T. I., Biochim. Biophys. Acta *155*, 626 (1968).
32. RajBhandary, U. L., Chang, S. H., Stuart, A., Faulkner, R. D., Hoskinson, R. M., and Khorana, H. G., Proc. Nat. Acad. Sci. (Wash.) *57*, 751 (1967).
33. Razin, A., Sedat, J. W., and Sinsheimer, R. L., J. Mol. Biol. *53*, 251 (1970).
34. Robins, M. J., Hall, R. H., and Thedford, R., Biochemistry *6*, 1837 (1967).
35. Roscoe, D. H., and Tucker, R. G., Biochem. Biophys. Res. Commun. *16*, 106 (1964).
36. Silagi, S., Cancer Res. *25*, 1446 (1965).
37. Sinsheimer, R. L., Science *120*, 551 (1954).
38. Smith, J. D., and Dunn, D. B., Biochim. Biophys. Acta *31*, 573 (1959).
39. Sober, H. A., and Harte, R. A. (Eds.), *Handbook of Biochemistry*, The Chemical Rubber Co., Cleveland, p. H-39 (1968).
40. Steiner, R. F., and Beers, R. F., *Polynucleotides*, Elsevier Publishing Co., Amsterdam (1961).
41. Takahashi, I., and Marmur, J., Nature *197*, 794 (1963).
42. Unger, G., and Venner, H., Z. Physiol. Chem. *344*, 280 (1966).
43. Vanyushin, B. F., and Belozersky, A. N., Dokl. Akad. Nau. SSSR *129*, 944 (1959).
44. Vanyushin, B. F., Belozersky, A. N., Kokurina, N. A., and Kadirova, D. X., Nature *218*, 1066 (1968).

45. Vanyushin, B. F., Tkacheva, S. G., and Belozersky, A. N., Nature 225, 948 (1970).
46. Volkin, E., J. Amer. Chem. Soc. 76, 5893 (1954).
47. Wyatt, G. R., Biochem, J., 48, 584 (1951).
48. Wyatt, G. R., and Cohen, S. S., Biochem, J., 55, 774 (1953).
49. Zachau, H. G., Dütting, D., and Feldmann, H., Hoppe-Seyler's Z. Physiol. Chem. 347, 212 (1966).

Chapter 2 MOLECULAR STRUCTURE OF THE NUCLEIC ACIDS

The simplified concept of RNA as a single-stranded molecule and DNA as a double-stranded molecule defines the major difference between the two types of nucleic acids, although there are exceptions to this rule in certain viruses. However, this simple distinction does not take into account the complex arrangement of the nucleic acids into the cell organelles or the secondary and tertiary structure of these molecules. With the development, in recent years, of sophisticated equipment and correspondingly sophisticated techniques for isolation, identification and characterization of the nucleic acids, particularly of various types of RNA molecules, it has been shown that earlier assumptions (that RNA is cytoplasmic and DNA is confined to the nucleus, for example) are no longer valid. Both major types of nucleic acid have been shown to be widespread throughout the cell and to occur as various distinct types both in terms of function and structure. Owing to their stability and relative ease of isolation some types of nucleic acid have been studied extensively. Other types which occur in very small amounts or are unstable have received very little study, and reports of new types of nucleic acid appear regularly.

OCCURRENCE, COMPOSITION AND PRIMARY STRUCTURE

Ribonucleic Acids

Ribonucleic acids occur in animals, plants, bacteria and viruses. RNA viruses, in which RNA is the sole nucleic acid component, have been found in animal, plant and bacterial hosts. In eukaryotic cells RNA occurs both in the nucleus and cytoplasm and has been shown to be a constituent of various cellular structures such as chloroplasts and mitochondria. In animals, plants and bacteria the major types of RNA that have been described are ribosomal RNA, messenger RNA, transfer RNA and 5 s RNA. More recently a 7 s RNA[122] and a series of RNA's from 4 s to 10 s[27,153,178,189,190,272] have been described. In the nuclei of plants and animals high molecular weight RNA's have been described;[26,191-193,224,225,269] these are precursors of ribosomal RNA and are 45 s in size.

Ribosomal RNA

Ribosomal RNA is the most abundant and one of the most stable RNA's in most tissues. It occurs in the cell in combination with protein in the form of discrete particles called ribosomes. In mammalian and plant cells the ribosomes are found free in the cell sap or in clusters joined by messenger RNA, forming polyribosome structures,[228,267] or attached in polyribosome formation to the endoplasmic reticulum, a highly organized filamentous structural entity in the cell. Their organization in the cell depends on the type of cell under investigation. Ribosomes also occur in chloroplasts,[20] in mitochondria,[145] and have been reported in the nuclei of some cells.[198] The latter report, however, is currently under debate. Bacteria ribosomes, active in protein synthesis, occur in polysome formation,[96,210,220] but bacterial cells have no endoplasmic reticulum, no mitochondria and no nucleus. During their biological functioning ribosomes are associated with messenger RNA and transfer RNA in the cell. However, a large amount of evidence shows that these two types of RNA are completely separate entities from ribosomal RNA.

A ribosome is composed of two subunits which can be dissociated by reducing the magnesium concentration of the medium.[40,258,261] Each subunit contains a molecule of RNA associated with protein. Noll[242] has suggested that there are three classes of ribosomes in nature: animal ribosomes 80 s in size, the subunits of which contain 29 s and 18 s molecules of ribosomal RNA, 80 s ribosomes of plant cells, which contain 25 s and 16 s ribosomal RNA in their subunits, and 70 s bacterial ribosomes, which contain 23 s and 16 s ribosomal RNA in their subunits. In the third group, the bacterial ribosomes, Noll includes the chloroplast ribosomes; the mitochondrial ribosomes, since these are similar to the bacterial ribosomes, would also fit in this classification.[145]

When 70 s bacterial ribosomes dissociate, they give rise to two subunits, a 50 s subunit and a 30 s subunit.[258] Associated with the 50 s subunit is a small RNA molecule, the 5 s RNA molecule,[213] which will be discussed separately. The 50 s subunit has a particle weight of 1.8×10^6 daltons,[258] and its 23 s RNA component a molecular weight of 1.1×10^6 daltons.[146,171,239] Approximately 35 protein molecules are associated with the 50 s subunit.[184,260] The 30 s subunit has a particle weight of 0.85×10^6 daltons,[258] and the 16 s RNA component a molecular weight of 0.55×10^6 daltons.[146,171,239] There are approximately 20 protein molecules in the

24 / MOLECULAR STRUCTURE OF THE NUCLEIC ACIDS

30 s subunit.[147,184,260] The proportion of RNA to protein in both subunits is the same, 64 per cent RNA and 36 per cent protein,[258] as well as a small amount of polyamines.[230]

Ribosomal RNA shows the characteristic asymmetric base composition of almost all RNA's with a high guanine and low cytosine content.[185,231] The base compositions of the 16 s and 23 s ribosomal RNA molecules are similar but show significant differences[170] in terms of both nucleotide distribution[3,217] and the 5' terminal nucleotide sequences.[244,249] As mentioned in Chapter 1, ribosomal RNA contains some methylated bases, and some of the

Table 2–1. The 5'-Terminal Nucleotide Sequences of the 23 s and 16 s Ribosomal RNA's from Different Bacterial Species*

Organisms	23 s rRNA	16 s rRNA
Bacillus cereus	pUpXpXpXpGp——	pUpXpXpXpGp——
Bacillus subtilis	pUpXpXpXpGp——	pUpXpXpXpGp——
Bacillus stearo-thermophilus	pUpXpXpXpGp——	pUpXpXpXpGp——
Escherichia coli B	pGpGpUp ——	pApApApUpGp ——
Escherichia coli A19	pGpGpUp ——	pApApApUpGp ——
Sarcina lutea	pUpUpGp ——	pUpXpXpXpGp ——

* From Osawa, S., Ann. Rev. Biochem. *37*, 109 (1968).

ribose moieties are methylated in the 2' hydroxyl position. The methyl derivatives occur in both the 16 s and 23 s RNA, but the distribution is different in the two species.[110,183] The presence of methyl groups in the 2' hydroxyl position of ribose has been used as a method for examining nucleotide distributions in various ribosomal RNA's based on the alkali stability of these residues.[74,150,154,183,265] Sequence studies on ribosomal RNA are still in the early stages mainly because of the huge size of the molecules involved, but work on the 16 s and 23 s RNA and on some specific areas of these molecules has already provided some data.[75-78] The terminal sequences, particularly the 5' terminals, of a number of ribosomal RNA's have been studied, and a comparison of the termini from a number of different organisms has been reported by Sugiura and Takanami[244,249] (Table 2–1).

Messenger RNA

Messenger RNA carries the genetic information coded in the sequence of nucleotides from the DNA for ultimate translation into protein. In the majority of tissues it is short-lived—estimates vary from a few seconds to a few hours to a few days. In bacterial systems the messenger is short-lived, the current concept being that it is transcribed (that is, synthesized) from the DNA, that the translation of its message into protein occurs immediately while it is still attached to the DNA molecule, and that before the end of its synthesis the first part of the message undergoes degradation.[148] In such a metabolically fast system it is difficult to isolate a defined sample, and since the amount of messenger RNA in any cell is very small in terms of the total RNA of the cell, sufficient messenger RNA for good chemical and physical characterization has not yet been isolated from most systems. Thus, knowledge of the composition and structure of messenger RNA must be inferred from metabolic studies.

Messenger RNA is found in all tissues that synthesize protein and thus is a component of all animal, plant and bacterial cells. Like all RNA's it is considered to be single-stranded, composed of a single, uninterrupted chain of repeating nucleotide sugar phosphate residues. According to the currently held theory of protein synthesis, messenger RNA should be unassociated with any structural protein in bacteria and viruses. In eukaryotic systems, in which transcription occurs in the nucleus, the messenger RNA must be transported from the nucleus to the cytoplasm. Studies on this translocation process indicate that messenger RNA packaged together with protein may move from the nucleus to the cytoplasm into ribonucleoprotein particles which have been called informosomes.[114,115,182,233,234] From radioisotope incorporation studies it appears that messenger RNA contains no unusual bases,[240] and since it is theorized that in bacteria its message is translated into protein before its synthesis is complete, any modification of the bases prior to the onset of protein synthesis would have to be extremely rapid and probably associated with RNA polymerase, the enzyme which synthesizes RNA. This is most unlikely, particularly since the methylating enzymes for other types of RNA have been isolated and most modifications have been shown to occur at the polymer level.[26,235] The molecular weight of messenger RNA will vary according to the type of protein for which it codes. In instances in which the synthesis of certain proteins is linked, the linking effect, or polycystronic effect, occurs at the genetic level, and presumably the messenger RNA's will be the messenger for more than one polypeptide. Since protein synthesis has been reported to occur in the nucleus as well as in the cytoplasm of eukaryotic cells[198] and in chloroplasts[104,127] and mitochondria[104,127] messenger RNA should also be present in these cellular organelles. The majority of studies on messenger RNA in systems in which messenger RNA is required *in vitro* have utilized either synthetic polyribonucleotides (see Chapter 3) or viral RNA, which will function in place of the regular messenger RNA because viral RNA is the viral messenger RNA.

Transfer RNA

Transfer RNA is the smallest RNA molecule found in the cell and occurs in relatively high proportion, approximately 10 per cent in most tissues. It is found in all animal, plant and bacterial cells, and in eukaryotic organisms it has been reported in mitochondria.[7,8] The molecule is stable and is unassociated with protein in its structure. As its name implies, its function in protein synthesis is to transfer the amino acid from free solution into its proper position in the polypeptide chain. It is the original adaptor molecule, postulated by Crick in his adaptor hypothesis,[55] and was first isolated by Hoagland and Zamecnik in 1958.[121] When first isolated it was called soluble RNA since it was found in the soluble portion (centrifugal fraction) of the cell, and this terminology is still used occasionally. The molecule is of small size—70 to 80 nucleotides in length, depending on the particular transfer RNA—with a molecular weight of 25,000 daltons, and is by far the most widely studied of all RNA molecules. In recent years extensive fractionation studies have been undertaken, and it has been shown that there exists a large number of transfer RNA's, specific to specific amino acids, and that there is more than one specific transfer RNA for some amino acids.[39,132,174]

In Chapter 1 the extensive number of minor components which occur in transfer RNA were discussed. Among the most interesting are the nucleoside pseudouridine, the sulfur-containing modified bases and the cytokinin-type bases. Cytokinins are plant hormones, and the discovery of these as minor base components in the transfer RNA in both plant and bacterial species is unusual. Whether their hormonal activity has any functional significance in transfer RNA is as yet unknown. Since the initial total sequence of a yeast phenylalanine transfer RNA was described by Holley in 1965,[123] a relatively large number of transfer RNA's have been sequenced completely. A partial list of the transfer RNA's sequenced to date is presented in Table 2–2. It is evident from the list that the transfer RNA's from different species coding for the same amino acid and different transfer RNA's from the same species coding for the same amino acid have been sequenced. In all these cases there are minor changes in the base sequence for one amino acid, showing that species specificity does exist and that within one species there is degeneracy even at the transfer RNA level. The sequencing of transfer RNA's has been aided considerably by the paper chromatographic and electrophoretic techniques developed in Sanger's laboratory,[24] and it is now theoretically possible to sequence a transfer RNA in approximately a three week period.

The availability of sequences of transfer RNA's has led directly to examination of the chemical properties of transfer RNA in relation to structure and function. The molecule is particularly adaptable to such studies because of the high content of minor base components which can be treated selectively. For example, inosine and pseudouridine can be selectively cyanoethylated,[286] and transfer RNA molecules can be split into half molecules by chemical or enzymatic means.[123,126,195] Recently transfer RNA has been crystallized,[43,108,138] and the first X-ray diffraction studies on single crystals of transfer RNA have been reported.[53,133,288] Sequences of all transfer RNA molecules examined so far exhibit a certain number of sequence similarities.[164,236] The majority of transfer RNA's have a guanine at the 5' end, although, for example, in yeast (both bakers' and *Torulopsis*

COMPOSITION AND PRIMARY STRUCTURE / 27

Table 2-2. Complete Sequences of Transfer RNA's

Transfer RNA	Source	Reference
Alanine I	Yeast	123, 169
Phenylalanine	Yeast	203, 205
Serine I	Yeast	289
Serine II	Yeast	289
Tyrosine	Yeast	165
Valine	Yeast	6
Isoleucine	*Torulopsis utilis*	250
Tyrosine	*Torulopsis utilis*	109
Valine	*Torulopsis utilis*	175
Methionine	E. coli	48
N-formylmethionine	E. coli	67–69
Phenylalanine	E. coli	9, 262
Tyrosine II	E. coli	65, 102, 204
Tyrosine (Su_{III}^+)	E. coli	102
Valine	E. coli	283
Phenylalanine	Wheat germ	70, 71
Serine	Rat liver	237

utilis) tyrosine transfer RNA the 5' terminus is a cytosine. Nucleotides 55 to 65 include a triplet T-ψ-C in all transfer RNA's, and in a large number of transfer RNA's the cytosine of the triplet has guanine as the 3' adjacent neighbor. Nucleotides 15 to 25 include all the dihydrouracil bases in each molecule. The similarities in the sequences have helped to establish some of the structure-function relationships in transfer RNA.

5 s RNA

5 s RNA was originally described by Rosset and Monier[213] and has since been shown to be an integral structural component of ribosomes, associated with the 50 s subunit.[90,214] It is a low molecular weight RNA of approximately 120 nucleotides. The complete nucleotide sequences of a number of 5 s RNA molecules have been reported.[24,81] Brownlee, Sanger and Barrell have shown that in *E. coli* there are at least two different 5 s RNA molecules which differ only in one nucleotide position.[24] A G in position 13 is changed to a U in one strain of *E. coli*, and a C in position 12 is changed to an A in another strain. 5 s RNA bears no relation to tranfer RNA. It contains no minor bases and does not have the biological function of transfer RNA. No definitive function has as yet been assigned to 5 s RNA, but it has been postulated as a structural link between the ribosome subunits.[51,176] Sequences of 5 s RNA from bacterial and mammalian cells, although showing a certain amount of homology, are very different.[24,81] Like transfer RNA, 5 s RNA is unassociated with protein.

Minor Uncharacterized RNA Species

The development of sophisticated isolation techniques for macromolecules, which led to the careful characterization of the major RNA species, has resulted in a number of reports on minor RNA species. Careful isolation of RNA from mammalian nuclei results in large molecules of RNA which have been shown to be precursors of ribosomal RNA.[26,191–193,224,225,269] Similar studies have also resulted in reports of small RNA species of 7 to 10 s RNA which are deleted from these large precursor molecules but which are not ribosomal RNA, 5 s RNA or transfer RNA precursors.[27,122,153,178,189,190,272] These molecules have not been isolated in quantity nor have they been fully characterized. The major component is a 7 s RNA molecule, which in rat liver is rich in uridylic acid.[122] The 7 s RNA that occurs in higher proportions in Novikoff hepatoma cells is not a degradation product of the high molecular weight ribosomal RNA precursor and is not a precursor of 5 s RNA.[178]

In bacterial cells a species of RNA of approximately the same size as 7 s RNA has been reported which is distinct from transfer RNA and 5 s RNA.[100] All RNA's found associated with chloroplasts or mitochondria are counterparts of the RNA molecules found elsewhere in the cells, such as ribosomal RNA, transfer RNA and messenger RNA.

Viral RNA

Viruses in which the only nucleic acid component is RNA have been found in bacteria, plants and animals. This RNA is unusual in that it is the genetic material of these organisms and in some of the viruses functions as messenger RNA. Single-stranded viral RNA has been used widely as a substitute for messenger RNA in experiments on protein synthesis. Viral RNA is found in both the single-stranded form and in some mammalian viruses as a double-stranded nucleic acid molecule. Some of the mammalian viruses which are oncogenic have recently been shown to act as templates for the synthesis of DNA, which in turn is the template for synthesis of more viral RNA so that the RNA of the virus has a peculiar genetic type function.

The structure of the viruses—both RNA and DNA viruses—is essentially a nucleic acid core surrounded by a number of protein subunits. Accurate models of a number of viruses have been constructed.

BACTERIAL VIRUSES. A large number of bacterial RNA viruses have been isolated, but structural studies have involved primarily R17, MS2 and Qβ. The RNA of the bacterial viruses contains no unusual bases.[98] The RNA of the bacterial viruses is single-stranded and there is one RNA molecule per virus particle (virion). Sequence studies have been started on these viral RNA's, and sequences of nucleotides from the 3' and 5' ends of all three have been determined.[98] The 5' end of the RNA in each virus carries a terminal nucleoside triphosphate.[60,61,212] The sequence studies have revealed that certain regions of the RNA are folded or looped into base-paired regions.[49,135] This raises the interesting question of what happens to

COMPOSITION AND PRIMARY STRUCTURE / 29

this secondary structure during translation. Presumably it must be unfolded by the ribosome for the amino acid code to be read.

PLANT AND ANIMAL VIRUSES. Early work on viruses focused on the tobacco mosaic virus, which was the first virus discovered,[128] the first virus purified[238] and the first virus from which infectious nucleic acid was isolated.[82,94,95] The RNA in tobacco mosaic virus is single-stranded, and a model of the RNA core with the proteins attached around the outside has been constructed[142] (Figure 2-1). The development of tissue culture techniques has aided the study of mammalian viruses, some of which are very complex. Some viruses, such as the reovirus, contain more than one molecular species of RNA, and the RNA is double-stranded.[101,152] Sequence studies on these larger RNA molecules have not progressed as far as those on the bacterial viruses. The isolation of double-stranded RNA from a mammalian virus is interesting because for many years it was speculated that double-stranded RNA could not exist or, if it did, would not be a good genetic material. This latter argument may still hold for a complex mammalian system, but in a simple viral system which does not require the genetic integrity of a highly ordered species this is of less consequence. Also, the reproductive cycles of single-stranded plant and mammalian RNA viruses and particularly the bacterial RNA viruses have been shown to go through a double-stranded RNA intermediate form.[275-277]

Deoxyribonucleic Acids

The majority of DNA found in eukaryotic organisms is located in the nucleus and associated with the chromosomes. In bacteria the majority of DNA is found associated in a distinct chromosomal mass, bacteria not having any nucleus bounded by a nuclear membrane. However, in the past few years DNA has also been located in a number of extra-nuclear bodies. In eukaryotic cells it is found in mitochondria and in chloroplasts,[104] and in bacteria separate molecules of DNA, called episomal DNA's have been found unassociated with the chromosomal elements.[66] The quantity of these different types of DNA is small compared to the "nuclear" DNA, but in a number of instances they have been fairly well characterized. These findings have destroyed the concept that DNA is located entirely in the nucleus.

Many viruses—both animal and plant viruses as well as the bacterial viruses or bacteriophages—have DNA as their sole nucleic acid constituent. The DNA viruses provide a readily available source of easily purified DNA which is not contaminated with RNA and relatively easy to dissociate from the viral protein components, consequently, a large amount of the work on DNA has focused on the DNA viruses, particularly bacteriophages.[32] Investigation of DNA-containing bacterial viruses has led to the discovery of two groups of bacteriophages which are unusual because they contain single-stranded DNA. The two groups are the small, filamentous and spherical

30 MOLECULAR STRUCTURE OF THE NUCLEIC ACIDS

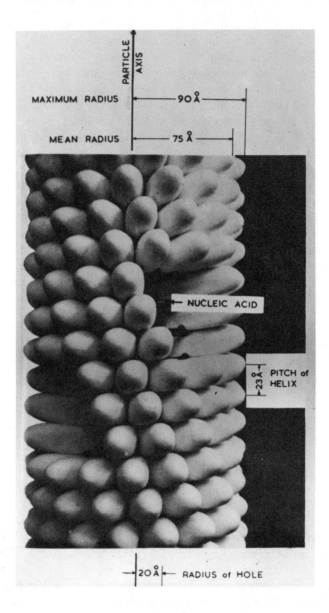

a

Figure 2-1. A model of tobacco mosaic virus, based mainly on X-ray diffraction studies. About one-tenth of the total length of the virus particle is shown.

The virus protein is in the form of a large number of small equivalent subunits in a helical arrangement about the particle axis. The structure repeats every 69 Å in the axial direction. A repeat unit comprises 49 subunits distributed over three turns of the helix, which has a pitch of 23 Å. The helical array of subunits has a hollow core of 35 to 40 Å diameter.

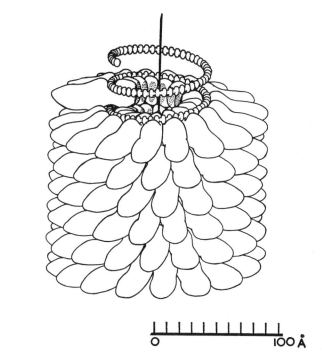

b

Figure 2-1 *Continued.* A schematic diagram of the model of tobacco mosaic virus shown in Figure 2-1, with part of the protein subunit structure omitted to reveal the organization of the viral ribonucleic acid molecule in the core of the model.

There are three nucleoticles per protein subunit, or 49 per turn of the helix, spaced about 5 Å apart. [From Klug, A., and Caspar, D. L. D., Adv. in Virus Res. 7, 225 (1960).]

ϕX174 types. Thus, a second common conception—that all DNA is double-stranded and has to be double-stranded in order to be the genetic material—is no longer valid. However, these cases are unusual, and the majority of the DNA in organisms is double-stranded.

Animal and Plant DNA

NUCLEAR DNA. As stated earlier, the majority of DNA in animal and plant cells is located in the nucleus. DNA is a high molecular weight molecule, estimates in size ranging up to 300×10^6 daltons. In the nucleus the DNA is organized into the chromosomal apparatus and is associated with protein and some RNA. The majority of the chromosomal proteins are histones, which are highly basic proteins rich in the amino acids arginine, lysine and glycine. A lesser amount of protein—acidic protein—is also present in the chromosomal structure.[113,120] In animals and plants the DNA is heterogeneous; that is, there are many different molecules of DNA with different sequences organized into the chromosomes. How the DNA molecules are arranged into the chromosome is still unclear; the subject has been a matter of much speculation and a number of reviews have appeared on the subject.[12,113,252] Minor bases have been found in all animal and plant DNA's so far examined, and proportions vary according to the source of the DNA. However, plant DNA's seem to be particularly rich in 5-methylcytosine, and up to 6 per cent of the cytosine can be present as 5-methylcytosine.[263,282] All plant and animal DNA's obey the base-pairing rule of Chargaff: A = T and G = C. Because of the large size and heterogeneity of plant and animal DNA's, little sequence work has been carried out. That sequence work which has been carried out will be described in a later section.

SATELLITE DNA. When some plant and animal DNA's are isolated carefully and the DNA centrifuged in a cesium chloride density gradient, a portion of the DNA bands at a density different from the main DNA component. Usually the density of this material is lighter but not exclusively so. This material has been named satellite DNA and is nuclear in origin. Satellite DNA was first isolated from the crab *Cancer borealis*,[243] and later from a number of other species.[139] Crab satellite DNA is still the most unusual that has been discovered. It comprises, in *Cancer borealis*, 30 per cent of the total DNA of the cell[229] and is approximately 97 per cent AT base-paired.[247] No other DNA has such an unusual base composition or is present in such large quantities relative to the main DNA component.

The isolation of satellite DNA from the crab stimulated a search for satellite DNA's in other organisms since this provided a readily available source of an unusual DNA component for study. This is particularly important since no satisfactory methods of fractionating DNA have yet been developed, mainly because DNA molecules from the same species differ not in size but in sequence of nucleotides. Search for satellite DNA's in other tissues have shown them to be present in mouse, guinea pig and human tissues, and in calf thymus DNA there are at least two satellite DNA's.[197,199] Satellite DNA has also been reported in bacterial preparations, but this discovery will be discussed separately.

The satellite DNA from the crab is a true satellite, differing grossly in composition from the main band DNA and separable by centrifugation from high molecular weight preparations of the total DNA. However, in order to obtain a good separation of most mammalian satellite DNA's by centrifugation, it is necessary to shear degrade or denature and renature the DNA. Mouse satellite DNA has been shown to contain 5-methylcytosine,[216] and the complementary strands of the preparation have been separated by centrifugation.[80] When exposed to heat (100°C) and then slowly cooled, satellite DNA renatures;[22] that is, the changes in conformation (separation of the two strands) due to heating are reversed and the molecule assumes its normal double-stranded configuration. Generally, because of its heterogeneity, mammalian DNA does not renature after heating and slow cooling. This and other evidence suggests that satellite DNA contains highly repetitive sequences of DNA thought to be approximately 300 to 400 nucleotides long.[266]

Satellite DNA has been isolated by other techniques such as hydroxyapatite column chromatography[13] and has been demonstrated by the DNA agar binding technique.[266] Several laboratories have used various terminologies for this component such as stabile, labile, fast renaturing, satellite, and so on. McLaren and Walker[168] have recently examined the various techniques used for isolation of various fractions of DNA from mouse tissue and have shown that "stabile" DNA corresponds to the "fast renaturing" DNA described by Britton and Waring[22] and is very similar to the AT-rich minor component, or satellite DNA, first demonstrated by cesium chloride density gradient centrifugation.[139] That there is confusion in terminology and that satellite DNA from mammalian tissues is best isolated following shear degradation and denaturation supports the suggestion by Bernardi[13] that what has been isolated as satellite DNA is a series of repeating nucleotide sequences which are present in DNA molecules but which may not be present as separate molecular entities. Recently Brahic, in Fraser's laboratory, has isolated from mouse DNA a rapidly renaturing fraction with many properties which indicate that it is composed of repeating sequences and has shown that this is equally distributed between the main band and the satellite band of mouse DNA following cesium chloride density gradient centrifugation.[19] This fraction was isolated by digestion of the unrenatured DNA (the DNA was denatured by heat, then renatured by slow cooling) by *Neuropora crassa* nuclease. These experiments support the idea that the repeating nucleotide sequences are distributed throughout the total DNA of the mouse. In plants, satellite DNA has been isolated from pumpkin[166] and from a yeast cytoplasmic "petit" mutant.[14]

CYTOPLASMIC DNA. DNA has been demonstrated in mitochondria and in chloroplasts,[104] and in mitochondria has been shown to be a completely separate entity from nuclear (and satellite) DNA, its properties being more similar to bacterial DNA than to mammalian DNA. Mitochondrial DNA molecules have been shown to be circular in a number of organisms studied—yeast,[5] HeLa cells,[125] L-cells,[181] human leucocytes[44] and unfertilized sea urchin eggs.[196] Some of the circular molecules in certain cell types occur in catenated forms and as circular dimers. This has generated numerous speculations concerning the synthesis of mitochondrial DNA.[44,125,181,196] Mitochondria are thus unusual in that they are almost cells within cells; they contain DNA, ribosomes, transfer RNA, messenger RNA[4] and the complete genetic and protein synthesizing system normally associated with a completely functioning separate cell. This fact in turn has led to speculations concerning the origin of mitochrondria.[4,45] Chloroplasts, the equivalent structures to mitochondria in plants, have not been studied as much as mitochondria owing to the difficulty in isolating the nucleic acids. However, those studies which have been made on chloroplasts indicate a close similarity to mitochondria.[127] Nass has examined the intra-mitochondrial distribution of DNA in L-cells; she has shown that DNA is contained in certain regions called nucleoids and has estimated an average of 2 to 6 molecules of DNA per mitochondrion.[181] Approximately 80 per cent of the DNA molecules were associated with the mitochondrial membranes.

Bacterial DNA

Bacterial DNA is double-stranded and is associated in a chromosome-type structure in the cell; it differs, however, from mammalian DNA in that the protein associated with the DNA is not histone protein. Careful isolation of a number of bacterial DNA's has indicated that a single bacterium contains a single molecule of DNA, and Cairns has shown that in *E. coli* this is a circular molecule.[30] The technique used to show the circularity of *E. coli* DNA involved radioautography of gently lysed *E. coli* cells prepared for electron microscopy. Because of the difficulties inherent in the technique, few other bacteria have been examined.[15,162] The molecular weight of *E. coli* DNA has been determined to be 2.2×10^9 daltons.[31]

Satellite DNA has been described in the bacteria *Bacillus cereus*,[66] *Rhodospseudomonas spheroides*[93] and in the extremely halophilic bacteria.[177] Satellite DNA's have not been examined widely in bacteria because of the relatively few reports of their occurrence. Of more interest to nucleic acid chemists has been the number of reports of minute DNA molecules also present in bacteria.[11,52,99,133,157,158,215] The majority of these minute DNA's have been shown to be circular.[11,52,99,133,157,158,215] Some of these forms of DNA have been shown to be episomal DNA, some have been shown to be hereditary determinants which are transiently extra-chromosomal,[131] and others have been shown to be bacterial plasmids, extra-chromosomal genetic elements that exist solely as autonomous entities.[156] The episomal pieces of DNA are considered to be specifically susceptible to alterations by certain environmental agents, such as acridine dyes, which can readily cause mutations.[129] Some of the episomes are sex factors, others are colicinogenic factors.[11,118] These are extra-chromosomal genetic elements that are responsible for the production of specific antibiotic materials—colicins—in bacteria that confer upon their host immunity to the antibiotic which they determine. Other extra-chromosomal DNA's in bacteria have not been assigned any particular biological function. Extra-chromosomal elements have been described in *E. coli*,[11,52,133] *Shigella dysenteria*,[215] *Micrococcus lysodiecticus*[157,158] and a number of other organisms.

Viral DNA

Mammalian, plant and bacterial viruses occur in which DNA is the sole nucleic acid component. Like the RNA viruses they have been studied extensively owing to the relative ease of purification of high molecular weight homogeneous DNA, since each viral particle contains only one DNA molecule and because, since RNA is not present in the virus, it is not necessary to purify from this component. Studies of the genetics and biochemistry of bacteriophages and the recognition of these relatively simple parasites as excellent models for these studies have largely been responsible for that area of biology known as molecular biology. An excellent *Festschrift* to Delbrück, one of the pioneers in this field, was published in 1966.[32]

Most of the studies on bacterial viruses have centered on the T series of coliphages, of which there are two groups, the T-even (T2, T4, T6) and the T-odd (T1, T3, T5, T7). The most comprehensive genetic information has been accumulated for coliphage λ. The bacteriophages which contain single-stranded DNA have been studied extensively also from the genetic point of view, and bacteriophages φX174 and S13 have been shown to be comprised of eight genes[134] and to go through a double-stranded replicative intermediate during synthesis of progeny phage.[227,253] Some of the most extensive sequence work performed on DNA's has been on the bacteriophages λ,[180,280] T7,[179] φX174[57,107,221] and S13.[37]

SEQUENCE STUDIES ON THE NUCLEIC ACIDS

Elucidation of the complete structure of a polymeric molecule depends, to a certain extent, on knowledge of the complete sequence of the component polymerized basic units. Overall three-dimensional and secondary-structural configurations do not absolutely require knowledge of the primary structure in detail, but the final minute details do require complete knowledge of sequence. The nucleic acids are unusual in that sequence information is minimal, and yet many studies and large amounts of information have been gathered on the secondary and three-dimensional configurations of the nucleic acids. Sequence studies on the ribonucleic acids are more advanced than those on the deoxyribonucleic acids for two major reasons—first, the availability of specific enzymes for degradation of ribonucleic acids, enzymes which hydrolyze RNA at specific base sequences, and second, the availability of highly purified, low molecular weight ribonucleic acids. Compared to the RNA molecules investigated, sequence work on DNA has been confined to certain relatively specific chemical degradative techniques on relatively high molecular weight molecules.

Ribonucleic Acids

The first RNA molecule to be sequenced, by Holley in 1965,[123] was phenylalanine transfer RNA. Holley utilized two specific enzymes—pancreatic ribonuclease, which hydrolyzes RNA at a 3' pyrimidine position, and T1 ribonuclease (takadiastase), which hydrolyzes RNA at the 3' guanine position. Hydrolysis by pancreatic ribonuclease provides a purine catalogue of a ribonucleic acid molecule, each of the oligonucleotides released being composed of purines with a pyrimidine at the 3' end. T1 ribonuclease provides a catalogue of mixed purine-pyrimidine oligonucleotides with guanine exclusively at the 3' termini. Holley utilized anion exchange column chromatography to separate the various oligonucleotide hydrolysis products of both enzymes and sequenced the oligonucleotides separated by using other enzymes such as snake venom phosphodiesterase, *E. coli* alkaline phosphomonoesterase and micrococcal nuclease. These techniques allowed him to obtain the total sequence of the two enzymatic catalogues and provided a large number of crossover areas, which in turn gave large sequence blocks. The final deciding technique in his sequence estimation depended upon the hydrolysis of transfer RNA by T1 ribonuclease at 0°C. Treatment of transfer RNA with the enzyme at 0°C results in hydrolysis of only one phosphodiester linkage in the molecule, cleaving the molecule into two large segments. Analysis of these two separated segments by his previous methods elucidated the complete sequence. Holley began his work in the late 1950's, following the first isolation of transfer RNA, and spent a number of years developing the countercurrent distribution technique for fractionation of transfer RNA's into homogeneous amino acid-accepting fractions.[2] Since his report of the original sequence, techniques have been refined, particularly in Sanger's laboratory where radioactively labeled RNA molecules and paper electrophoretic and chromatographic techniques have been used.[24] The chemical reactivity of some of the unusual bases in transfer RNA has also been exploited.[144,195,223,259]

In studies on 5 s RNA, which contains no unusual bases, chemical modification of the regular base components has also been carried out.[24] These modifications limit the specificity of the enzymes pancreatic ribonuclease and T1 ribonuclease, thus providing on hydrolysis different sets of oligonucleotides from the original digests. Two modification methods have been used for sequencing 5 s RNA. The first is the reaction of uridine and guanosine with a water soluble carbodi-imide derivative, N-cyclohexyl-N'-[β-morpholinyl-(4)-ethyl]-carbodi-imide-methyl-*p*-toluene sulphonate.[97] This derivative does not react with cytidine residues, and the substituted uridine is not susceptible to pancreatic ribonuclease hydrolysis so that specific splitting at cytidine residues can be achieved.[159] The second technique is methylation of RNA with dimethylsulfate, which preferentially methylates G residues to give 7-methylguanosine, which in turn are resistant to ribonuclease T1.[21,155] Using these two techniques Brownlee, Sanger and Barrell have sequenced the 5 s RNA from *E. coli*.[24]

Determination of the Sequence of E. coli 5 s RNA

In their determination of the nucleotide sequence of 5 s RNA, Brownlee, Sanger and Barrell[24] used ^{32}P-labeled 5 s RNA and two-dimensional paper fractionation methods to separate the oligonucleotides. Both of these methods increased the sensitivity of the assays and eliminated the need for huge quantities of starting material. The first analyses carried out were total pancreatic ribonuclease and ribonuclease T1 digestions. The oligonucleotides released (see Table 2-3) did not provide any overlapping sequence

Table 2–3. Products of Total Digestion of E. coli 5 s RNA by Pancreatic and T1 Ribonucleases*

RIBONUCLEASE T1		PANCREATIC RIBONUCLEASE	
Sequence	No. per Molecule	Sequence	No. per Molecule
G	11	C	19
CG	5	U	6
AG	2	AC	1
UG	4	GC	7
CCG	3.5	AU	2
AAG	1	GAC	1
CCAG	1	AGC	2
AAACG	1	GAAC	1
CCUG	1	GAAAC	1
UAG	3.5	GGC	2
AUG	1	AGGC	1
AACUG	1	GU	2.5
AACUCAG	1	GAU	1
ACCCCAUG	1	AGU	1
UCCCACCUG	1	GGU	3
UCUCCCCAUG	1	AGGGAAC	1
CCUUAG	0.5	AGAAGU	1
pUG	1	GAGAGU	1
CAU$_{OH}$	1	GGGGU	1
		pU	1
		AU$_{OH}$	1

* From Brownlee, G. G., Sanger, F., and Barrell, B. G., J. Mol. Biol. *34*, 379 (1968).

information when the products from pancreatic ribonuclease were compared with those from ribonuclease T1. To isolate larger degradation products they used the partial ribonuclease T1 digestion technique first used by Holley, et al.,[123] who showed that partial hydrolysis with ribonuclease T1 causes preferential hydrolysis at single-stranded regions of the molecule. Thus, the longer oligonucleotide degradation products should come from the more stable double-stranded regions. These longer oligonucleotides were separated by chromatography on DEAE paper, a homochromatography technique was used which utilizes a high concentration of a mixture of non-radioactive nucleotides to develop the chromatogram by displacing the radioactive nucleotides successively from the DEAE paper. This technique, used in combination with cellulose acetate ionophoresis, allowed separation of nucleotides up to 25 residues in length. The long oligonucleotides were then subjected to further digestion with ribonuclease A and T1 ribonuclease, and from this information the sequence of some of the oligonucleotides was deduced. Some examples of this are given in Table 2–4. A similar partial digestion with pancreatic ribonuclease was also performed with similar results. In order to complete information on the total sequence, partial digestion with spleen acid ribonuclease and chemical modification of the

Table 2-4. Products of Partial Ribonuclease T1 Digestion*

Products of Complete Digestion		Sequence of Oligonucleotide
Ribonuclease T1	Pancreatic Ribonuclease	
UCCCACCUG	GAC	UCCCACCUGACCCCAUG
ACCCCAUG	AU	
	AC	
	G	
AACUCAG	GAAC	CCGAACUCAG
CCG	AG	
AACUCAG	A_3	AACUCAGAAG
AAG	G_2	
	AAC	
AACUCAG	A_3	CCGAACUCAGAAG
AAG	G_2	
CCG	GAAC	
AACUG	AGGC	AACUGCCAGGCAU$_{OH}$
CCAG	GC	
CAU$_{OH}$	AAC	
G	AU$_{OH}$	
AACUG	GGAAC	GGAACUGCCAGGCAU$_{OH}$
CCAG	AGGC	
	GC	

* From Brownlee, G. G., Sanger, F., and Barrell, B. G., J. Mol. Biol. *34*, 379 (1968).

RNA followed by degradation with ribonuclease T1 were performed. The sequence of the oligonucleotides released with acid ribonuclease was determined following their complete digestion with T1 ribonuclease and pancreatic ribonuclease. When the 5 s RNA was methylated with dimethylsulfate and digested with ribonuclease T1, the partially methylated oligonucleotides were examined by digestion with pancreatic ribonuclease and by compositional (base) analysis. The results with the methylated RNA provided information on the sequence of certain oligonucleotides by comparison with the total end products of digestion and with other partial products.

An example of this is the oligonucleotide [U,(A,G)C]G, isolated from the methylated RNA. Four possible sequences are (1) UAGCG, (2) UGACG, (3) AGCUG and (4) GACUG. Comparing the T1 end products (Table 2-3), the oligonucleotides ACG, CUG and ACUG do not occur. Thus, sequences (2), (3) and (4) cannot occur. This indicates that sequence (1), UAGCG, is the correct sequence. A similar procedure was used with

Table 2-5.

1. pUGCCUGGCGGCCGUAG
2. GUAGCGCGGUGGUCCCACCUGACCCCAUGCCGAACUCAGAAGUGAAACGCCGUAG
3. GUAGCGCCGAUGGUAG
4. GUAGUGUGGGGUCUCCCAUGCGAGAGUAG
5. GUAGGGAACUGCCAGGCAU$_{OH}$

SEQUENCE STUDIES ON THE NUCLEIC ACIDS / 39

pUGCCUGGCGGCCGUAGCGCGGUGGUCCCACCUGA
CCCCAUGCCGAACUCAGAAGUGAAACGCCGUAGCGCCG
AUGGUAGUGUGGGGUCUCCCCAUGCGAGAGUAGGGAA
CUGCCAGGCAU$_{OH}$

Figure 2–2. Sequence of *E. coli* 5 s ribosomal RNA.

the carbodi-imide derivative-modified 5 s RNA oligonucleotides. Separation of these oligonucleotides on cellulose acetate at pH 3.5, followed by chromatography on DEAE paper at pH 1.9, showed that the blocked oligonucleotides were distributed to the positive side of the unmodified nucleotides in the first dimension, because of their extra positive charge at pH 3.5, and they showed greater mobility in the second dimension than the unmodified oligonucleotides because of the two additional positive charges at pH 1.9. Following removal of the carbodi-imide derivative group, the separated oligonucleotides were digested with pancreatic ribonuclease and ribonuclease T1, and from these digestions the structure deduced. The deduction of the final sequence was made by comparison of the sequences of the longer oligonucleotides derived from all of the accumulated results and was based on the identification of five long oligonucleotides (Table 2–5). The reasoning behind the allocation of the actual sequences from the experimental data is given in detail in the paper by Brownlee, Sanger and Barrell.[24] The complete sequence is shown in Figure 2–2, and the two alternative possible base paired structures suggested by these authors are shown in Figure 2–3. These

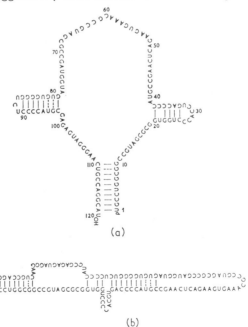

Figure 2–3. Suggested base-paired structures of *E. coli* 5 s RNA. A solid line indicates a standard G-C or A-U base pair and a dashed line a G-U pair. [From Brownlee, G. G., Sanger, F., and Barrell, B. G., J. Mol. Biol. *34*, 379 (1968).]

have been criticized by a number of workers because of other physical evidence which indicates that there is approximately 60 per cent hydrogen bonding in 5 s RNA. This is discussed later in the section on conformation.

Deoxyribonucleic Acids

The smallest DNA molecules isolated to date, such as polyoma virus, *E. coli* plasmid or the single-stranded DNA phages, contain 5000 to 5500 nucleotides. These molecules are large compared to the smaller RNA molecules, such as transfer RNA with 80 nucleotides or 5 s RNA with 120 nucleotides. Also, no deoxyribonucleases have been isolated to date which have base specificities similar to pancreatic ribonuclease or ribonuclease T1. Because of these two facts much less effort has been put into attempts to sequence DNA molecules than RNA molecules. Studies on the sequence of DNA have centred on two main areas—development of chemical techniques to selectively cleave the DNA molecule and specific modification of bases to allow study of base sequence by electron microscopy.

The electron microscopic studies have been carried out almost exclusively in Beer's laboratory. Beer has developed techniques for specific conversion of thymine to anionic osmate esters[119] and addition of acyl hydrazides to cytosine.[92] The technique depends upon the electron density of the modified bases, thymine osmate ester being highly so, but the resolution of the electronmicroscope presents problems. The unmodified parts of the DNA molecule are not electron dense; consequently, shadowing is difficult and the density areas of the electron dense atoms attached to the DNA molecule tend to overlap into each other, making interpretation of the electron micrographs extremely difficult.

Specific chemical modifications of DNA have been investigated in a number of laboratories. Some of the earliest studies developed from the original observation in Chargaff's laboratory[251] that the purine N-glycoside linkage is more labile to acid than the pyrimidine N-glycoside linkage. Partial hydrolysis of DNA by sulfuric acid removes the purines and hydrolyzes the adjacent sugar phosphodiester bonds, releasing pyrimidine oligonucleotides of the general formula $Py_n P_{n+1}$ from the DNA.[222] This technique was later modified by Burton and Petersen, who used formic acid and diphenylamine,[28] the formic acid for hydrolysis of the N-glycoside bond and the amine for attachment to the resultant aldehyde group at the C-1' position as an aid to β-elimination of the adjacent phosphate esters. The method has the advantage that the formic acid hydrolysis goes to completion but is less rigorous than the sulfuric acid method; it does not result in release of any pyrimidine bases by N-glycoside cleavage of the pyrimidine N-glycoside bonds with prolonged hydrolysis, a factor which occurs with the sulfuric acid hydrolysis.[28,137,232] Using the formic acid-diphenylamine degradative technique and separating on DEAE cellulose columns[38,194] the oligonucleotides released, first according to chain length and then according

SEQUENCE STUDIES ON THE NUCLEIC ACIDS / 41

to base composition, the method has recently been developed so that a complete pyrimidine catalogue can be obtained from a DNA of up to 50,000 base pairs.[37,57,179,180] By separating the complementary strands of double-stranded DNA,[124,246] or by using the replicative intermediate of single-stranded phages, the complementary catalogue of purine base pairs is obtained with this analysis procedure.[37,57] Therefore, this method gives a catalogue similar to that given by pancreatic ribonuclease for RNA.

Work in a number of laboratories was initiated following the publication of a method for obtaining a purine catalogue from DNA,[105,106] based on the hydrazinolysis of DNA to remove the pyrimidines and followed by alkaline degradation to give the purine oligonucleotide catalogue. Unfortunately, the hydrazinolysis is not as specific as the formic acid-diphenylamine hydrolysis for the pyrimidine catalogue, and the alkaline degradation does not give a completely quantitative hydrolysis of the apyrimidinic acid produced.[34,35,42,72,136] To overcome some of these difficulties, modification of the reaction has been introduced which involves the addition of benzaldehyde to produce an addition product following removal of the pyrimidines with hydrazine.[42] Vanyushin has recently shown that degradation of the resulting apyrimidinic acid, following benzaldehyde treatment, with analine results in a specific degradation to purine clusters.[264] Thus, two methods are available for looking at DNA from the point of view of pyrimidine and purine clusters.

Other chemical techniques investigated have been the selective degradation of pyrimidines by permanganate oxidation,[58,112] the reactivity of hydrogen peroxide and hydroxylamine with DNA,[206,207] the selective modification of cytidine with semicarbazide,[111] the oxidation of pyrimidines, particularly thymine, by osmium tetroxide,[29] the reactivity of 2,4-dinitrophenylhydrazine in the deamination reaction[187] and the reaction of formaldehyde and formamide with DNA.[161,167,257] Some of the reactions have been shown to be fairly specific, such as with semicarbazide, osmium tetroxide and permanganate, but they have not yet been developed fully as analytical tools. At the present time there are no means available for linking the pyrimidine catalogues with the purine catalogues to obtain a good overall sequence of a piece of DNA.

A large number of laboratories are involved in the search for specific degradative enzymes for DNA. A number of enzymes have been used for limited sequence analysis of DNA, particularly *E. coli* exonuclease III[208] and snake venom phosphodiesterase and polynucleotide kinase.[248] Using this technique Wu[280] has recently described the partial sequence of the 12 nucleotides at the cohesive ends of bacteriophage λ DNA. The enzyme polynucleotide kinase, which transfers a gamma-labeled ^{32}P atom to the 5′ hydroxy terminus of an oligonucleotide, has been used for both RNA's and DNA's and the 5′ terminal nucleotides of some bacteriophage DNA's sequenced.[248,273] Thus, apart from the cohesive ends of bacteriophage λ, DNA, no sequences have yet been published.

CONFIGURATION OF THE NUCLEIC ACID

Early studies on the structure of the nucleic acids consisted of examination of the component nucleosides and nucleotides by chemical methods. Many of the early chemical results have since been confirmed by physical methods. Studies on the configuration of the nucleic acids began with the early X-ray diffraction studies. Furberg[87,88] showed that the D-ribose in cytidine was in the furanose configuration, the sugar residue being located on N^1, and confirmed the β-configuration of the glycosidic linkage. The X-ray patterns indicated that the six atoms of the pyrimidine ring all lie in the same plane and that the glycosidic linkage lies in the same plane as the purine ring. Furberg also demonstrated that the sugar ring is slightly non-planar, the C-3' atom being out of the plane by nearly 0.5 Å. Finally he showed that the pyrimidine ring and the D-ribose ring systems are oriented almost perpendicular to each other. Furberg's X-ray analyses and conclusions were of fundamental importance in the later deduction of the structure of the nucleic acids. The nucleotide model[89] based on the X-ray data gave the length of the molecule as approximately 3.4 Å. Thus, when nucleotides are piled one on top of the other with the planes of the bases perpendicular to the long axis of the molecule and the planes of the sugar rings, as well as the phosphate bonds, approximately parallel the long axis of the model, most of the components lie in planes 3.4 Å apart.[89] This led to the first explanation of the strong 3.4 Å reflection in X-ray diffraction studies on the nucleic acids, particularly DNA.

Configuration of the Ribonucleic Acids

Information on the structure and configuration of the ribonucleic acids has expanded greatly in recent years, particularly because of the availability of the primary sequence of a number of transfer RNA's and the ability to obtain pure homogeneous species of different types of RNA molecules. However, for some species of RNA which have been difficult to isolate, information is virtually non-existent and speculation is based on theories of biological function. The majority of the information on configuration for those RNA's that have been studied is based on physicochemical evidence since X-ray diffraction of RNA's has not been particularly successful owing to the complexity and diffuseness of the patterns. The recent crystallization of transfer RNA should, however, enable investigators to produce interpretable X-ray diffraction studies on this species of RNA in the very near future.

Configuration of Ribosomal RNA

Investigations of the secondary structure of ribosomal RNA's by various chemical and physical methods have led to the point where models of ribosomes have been devised which accommodate fairly closely the chemical and physical parameters and provide guides for further studies in this area. Ribosomes are composed of three basic units: the 30 s subunit, the 50 s subunit and the 5 s RNA attached to the 50 s subunit. As described earlier in this chapter, the molecular weight of the RNA molecules in the subunits is known accurately, and the number of protein molecules associated with the 30 s subunit and the 50 s subunit have also been determined. The helix content of reticulocyte ribosomal RNA has been determined by a variety of physical methods,[103] and it has been concluded that between 60 and 80 per cent of each of the ribosomal RNA molecules is double-helical and that a limiting value for the helix length is about 4 to 10 base pairs. The methods used for this determination were hyperchromicity studies, spectrophotometric titration, optical rotatory dispersion and alkaline hydrolysis. It has been shown that about 20 to 30 per cent of the total reticulocyte ribosomal RNA can be removed from the ribosomes by enzymic hydrolysis without changing sedimentation properties or appearance in the electron microscope, and that the amount of RNA removed is not increased by further addition of ribonuclease up to more than 1 mg of ribonuclease per mg of ribosomes, by increasing the temperature from 0° to 30°C or by first converting the RNA moiety into a single-stranded denatured form.[50] The sites where the RNA is digested by the ribonuclease are widely distributed throughout the polynucleotide chains since small fragments of RNA are produced following digestion. From the 50 s subunit the ribonuclease digests approximately 40 sites per 4500 nucleotides, each site separated by approximately 400 nucleotides, and for the 30 s subunit there are 15 sensitive sites per 1500 nucleotides.

The reactivity of ribosomal RNA with formaldehyde has also been investigated.[50] The RNA of the ribosome undergoes a conformational change when ribosomes are treated with 8 per cent formaldehyde at 70°C for 10 minutes and then cooled to 20°C. After this treatment no double-helical character can be detected by optical rotatory dispersion or hyperchromicity studies, although the sedimentation coefficient and the morphology of the ribosome, as determined by electron microscopy, is unaltered. From these studies it is evident that the RNA moiety of the ribosome is freely accessible to formaldehyde. Cox and Bonanou have utilized these studies and correlated them with the electron microscopic studies of Bruskov and Kiselev,[25] together with much other information on the physical properties, such as the radius of gyration, the diameter of the ribosome in solution, the axial ratios and the X-ray diffraction studies by Langridge[151] and have suggested a model in which the ribosomal RNA molecule is arranged in a series of hairpin loops, each comprising seven base pairs 3 Å apart and nine unpaired residues.[51] The flexible region joining one hairpin loop to another was taken to be 5 to 10 residues long. The protein subunits fit between adjacent hairpin loops, such that the flexible region of RNA joining one subunit of protein to the next protein subunit, plus the hairpin loops, permit protein-protein interactions between adjacent protein subunits. In the total 50 s

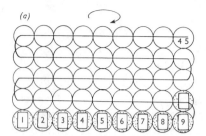
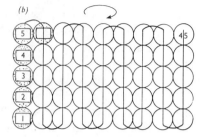

Figure 2-4. Diagram of sheets formed by folding a ribonucleoprotein thread constructed on the basis of the assumption that one protein subunit interacts with two hairpin loops. The thread is shown in its most compact form, comprising 45 repeating units and measures 80 Å × 200 Å × 360 Å. The arrow indicates the axis around which the sheet is folded to give a horseshoe. The protein subunits are represented by circles and the RNA double-helical segments by rectangles (for clarity only a few are shown). The direction of the ribosomal RNA chain is shown by the continuous line. The orientation of the double-helical segments with respect to the cylindrical axis of such a particle is not known. Two alternatives are (a), with five rows of nine subunits and the RNA double-helical segments parallel to the cylindrical axis, and (b), with nine rows of five subunits and the RNA double-helical segments perpendicular to the cylindrical axis. [From Cox, R. A., and Bonanou, S. A., Biochem. J. *114*, 769 (1969).]

subunit the arrangement of RNA hairpins and protein subunits can be considered as a sheet bent round to form a U-shape (Figure 2-4) so that the surface of the sheet exposes pieces of RNA hairpin and the inside forms a groove or cleft approximately 4 Å wide, as noted in the electron micrographs of Bruskov and Kiselev.[25] The 30 s subunit is composed of a similar bent sheet which fits like a cap on the open end of the U of the 50 s subunit (Figure 2-5). Cox and Bonanou[51] envisage the messenger RNA bound on the inside surface of the 30 s subunit parallel to the axis of the cylinder formed between the two units (Figure 2-6), with the growing polypeptide chain being extruded through the groove. They suggest also that the binding sites of both aminoacyl transfer RNA and the polypeptide transfer RNA lie within the groove. The dimensions of the proposed model agree with the dimensions described by Dibble and Dintzis[62] and Bruskov and Kiselev[25] from electron micrographs—180 Å × 180 Å × 240 Å. The axial ratio of the model also fits the axial ratio as determined by hydrodynamic methods (63). X-ray diffraction studies by Langridge[151] have shown a strong 45 to 50 Å reflection, which was interpreted as evidence for arrays of 4 or 5 double-helical segments of RNA separated by 45 to 50 Å. The proposed model has features consistent with this interpretation. Since a large portion of the RNA lies on the surface of the ribosome, these areas should be freely accessible to molecules small enough to penetrate the cleft; thus, the model provides an explanation for the accessibility of the nucleic acid moiety to cationic dyes, cations, and other small molecular species. The reactivity of the RNA with formaldehyde, causing conformational change from double-helical to single-stranded form without altering the sedimentation coefficient or the appearance of the electron micrographs, supports the

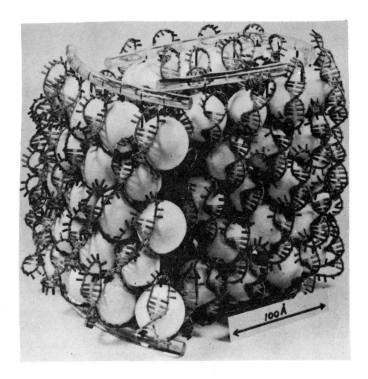

Figure 2-5. A model of a recticulocyte ribosome. Each sphere represents a unique protein. The hairpin loops of the RNA are not identical so that the surface of the model is heterogeneous. The smaller subparticle is fitted like a cap on to the larger horseshoe-shaped subparticle. The groove in the horseshoe has a minimum width of 20 to 30 Å. The messenger RNA would bind to the inside surface of the smaller subparticle possibly parallel to the axis of the cylinder, with the growing polypeptide chain being extruded through the groove. [From Cox, R. A., and Bonanou, S. A., Biochem. J. *114*, 769 (1969).]

concept that the RNA moiety is on the surface of the ribosome. The model predicts that the ribonuclease-sensitive areas would be the hairpin loops, about 20 for the smaller and about 45 for the larger subparticle, which correlates extremely well with the experimental evidence. The observation that a substantial part of the RNA is protected from enzymic hydrolysis agrees with the model which locates part of the RNA on the inner surface of a groove or hole within the ribosome (Figure 2-6).

Cox and Bonanou[51] suggest that the 5 s RNA which remains attached to the larger subparticle even after 20 per cent of the protein is removed, but is liberated when more protein is lost or on titration with EDTA, indicates that the 5 s RNA might be required to maintain the proposed horseshoe conformation. Another similar model has been proposed by Möller[176] which is also very similar to the electron micrograph model proposed by Bruskov and Kiselev.[25] Möller has considered the model in terms of the

46 / MOLECULAR STRUCTURE OF THE NUCLEIC ACIDS

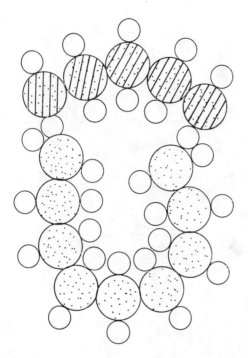

Figure 2–6. A schematic diagram of a cross section of the ribosome model shown in Figure 2–5. [From Cox, R. A., and Bonanou S. A., Biochem. J. *114*, 769 (1969).]

interaction between the ribosomal subunits, 5 s RNA, transfer RNA and messenger RNA during translation and proposed that in the 70 s ribosomal complex there is a total of some 40 substructures which are arranged cylindrically around a central axis shared by the two 50 s and 30 s subunits. He postulates approximately 13 substructures in the 30 s subunit and 26 in the 50 s subunit. Each structure comprises two predominantly double-helical RNA regions of 40 nucleotides and a guanine-rich region, which is not double-helical, of 30 nucleotides. This guanine-rich region interacts preferentially with protein subunits and is protected against nuclease. The link between the two double-helical regions is assumed to contain another 10 to 20 non-double-helical residues. Each substructure is thus composed of one protein and an RNA chain (spacer) of 120 to 130 nucleotide residues of which some 60 per cent are base-paired (Figure 2–7). His estimate of 13 proteins in the 30 s subunit and 26 in the 50 s subunit does not fit the experimental evidence very well since approximately 20 proteins have been isolated from 30 s subunits.[147,184,260] However, he suggests that some may be present in less than equimolar amounts and may reflect heterogeneity in the ribosome population. Möller's arrangement of particles, however, does fit the electron microscopic studies of Bruskov and Kiselev,[25] the extent of base pairing and helix content[103] and the susceptibility to RNase.[50] He postulates that the 50 s subunit is assembled similarly to the 30 s subunit, but is comprised of two concentric horseshoe configurations placed one above the other. His model of the 70 s ribosome has the overall dimensions of

CONFIGURATION OF THE NUCLEIC ACIDS / 47

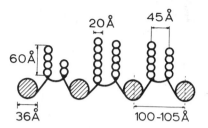

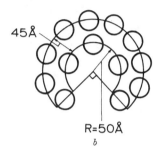

Figure 2-7. A model of the ribosome of *E. coli*. (*a*) A schematic diagram of the arrangement of the proteins and double-helical RNA regions in part of a ribonucleoprotein strand from an *E. coli* ribosome. (*b*) A schematic diagram of a cross section of the ribosome model along the central channel of the particle. The growing polypeptide chain is assumed to be present in the central channel of the 50 s subunit and thus protected from enzymatic digestion. [From Möller, W., Nature **222**, 979 (1969).]

250 Å × 200 Å, which approximates the measurements obtained from electronmicroscopic studies. He points out that the size of 5 s RNA—120 nucleotides—corresponds very closely to his suggested size of the spacer RNA's linking the two helical regions and on this basis assigns a spacer function to the 5 s RNA. In support of this argument he cites the amount of complementarity found between 5 s RNA and the suppressor III tyrosine transfer RNA, which is better than that expected for random sequences of similar lengths. However, this reasoning works against all other data, which indicate that 5 s RNA is a unique RNA molecule not related in structure or synthesis to transfer RNA, 16 s, or 23 s ribosomal RNA.[186] The advantage of these models is that they provide a basis for further investigations and by modification should allow construction of a final composite model of a ribosome similar to the models formulated for the RNA and DNA viruses.

Configuration of Messenger RNA

Information on the configuration of messenger RNA is negligible. From bacteria and mammalian systems no purified natural messenger RNA has been isolated in sufficient amount, state of purification or complete form to allow performance of any physicochemical studies. Because of this, conceptualization of the configuration of messenger RNA has depended on the acceptance of the biological theories of transcription and translation. It is thus assumed to be single-stranded and extended, since the ribosomal particles must be free to migrate along, or, alternatively, the messenger RNA is assumed to slide over or through the ribosomes, as the genetic message is read and translated into protein. Any secondary structure that

48 / MOLECULAR STRUCTURE OF THE NUCLEIC ACIDS

the messenger RNA may have must be readily alterable as the ribosomes move with respect to the messenger RNA itself. It has thus been assumed by most investigators that any helical formation or complex folding-back and hydrogen bonding between the bases of the messenger RNA to form areas of double-stranded RNA is unlikely. Also, messenger RNA should not be associated with any structural protein since this again would interfere with translation of the genetic message. Studies on viral messenger RNA, such as the messenger RNA's transcribed from T7 DNA which have been isolated,[245] support these inferences. However, sequence studies on the viral RNA R17[135] have revealed possible base paired, looped regions in this RNA, which has led to the suggestion that the loops must be opened for translation of the message. Hydrodynamic studies, particularly sedimentation analyses, have resulted in estimates of the size of messenger RNA's from 4 s to 16 s, depending on the source of the material. It has recently been suggested that as messenger RNA is transcribed in bacterial systems the ribosomes attach immediately and translate the message and that the degradation of the message starts before transcription is complete.[148] This explains many of the reasons why messenger RNA's have been extremely difficult to isolate from bacterial and mammalian systems. In mammalian systems the isolation of informosomes and the suggestion that these particles are the transport form of messenger RNA from the nucleus to the cytoplasm[114,115,182,233,234] raises important questions concerning the secondary and tertiary structure of the messenger RNA, with its interaction with the protein component of the informosomes, and how the protein is released when the message is to be translated. Studies on these particles hopefully will answer some of these questions.

Configuration of Transfer RNA

Holley, in 1956, in describing the first total sequence of a transfer RNA, also suggested a possible model for the secondary structure of transfer RNA's, which he termed the clover leaf model.[123] The base sequence of all subsequent transfer RNA's which have been determined all fit the base pairing clover leaf model,[236] and this is now the accepted secondary structure model for transfer RNA's. The model (Figure 2–8) consists of four arms—the amino acid arm with the CCA end group to which the amino acid becomes attached; a looped out arm called the TψC arm since the sequence TψC occurs in all transfer RNA's so far examined; the anticodon arm, which is also looped out, in which the triplet for read-out of the messenger RNA code occurs; and the dihydrouracil arm, a third looped out arm in which the modified dihydrouracil bases occur. In some transfer RNA's there is an extra arm between the anticodon loop and the TψC loop which varies in size.

In recent years, following the description of the sequences of a number of transfer RNA's (see Table 2–2), interest has centerd on developing models

CONFIGURATION OF THE NUCLEIC ACIDS / 49

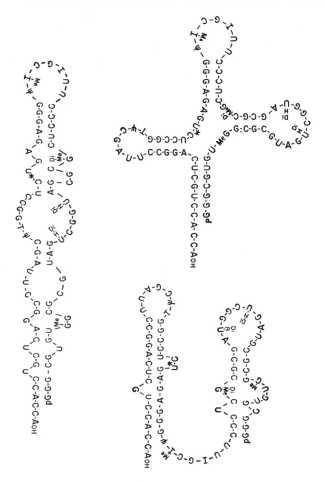

Figure 2-8. Suggested base-paired structures of alanine tRNA. [From Holley, R. W., Apgar, J., Everett, G. A., Madison, J. T., Marquisee, M., Merrill, S. H., Penswick, J. R., and Zamir, A., Science *147*, 1462 (1965).]

for the total configuration of transfer RNA. This model-building has received impetus from the crystallization of transfer RNA and the availability of some X-ray data, numerous physical measurements and precise chemical alterations which have been made on transfer RNA. One set of experiments aimed at determining the extent of hydrogen bonding in transfer RNA molecules was performed by Englander and Englander[73] and based on the hydrogen exchange principle first described by Printz and von Hippel[200]. Measurement of the hydrogen exchange behavior of mixed yeast transfer RNA allowed Englander and Englander to calculate approximately 77 hydrogen bonds per average 70 nucleotides in a transfer RNA molecule, indicating the involvement of approximately 82 per cent of the nucleotides in base pairing arrangements. They suggested that the non-hydrogen-bonded bases should loop out from the largely base-paired structure

50 / MOLECULAR STRUCTURE OF THE NUCLEIC ACIDS

except for the terminal CCA groups. These studies were performed before the data of Holley was available. Comparing the results of Englander and Englander with the clover leaf structure of Holley shows that there are 20 more hydrogen bonds in the transfer RNA molecule than shown in the clover leaf model.[123] Cramer, et al., in 1968,[54] showed that yeast phenylalanine transfer RNA reacts with monoperphthalic acid to a limited extent and that during the reaction 22 per cent of the adenosine residues are oxidized. When total yeast transfer RNA is oxidized, 27 per cent of the adenosine residues react, and when total *E. coli* transfer RNA is reacted, 36 per cent of the adenosine residues are oxidized. On the basis of these results and the splitting pattern of the oxidized material by pancreatic RNase, Cramer and his colleagues suggested that in total yeast transfer RNA two adenosine residues are involved in a structure which is lost at temperatures lower than 40°C or during the oxidation step. They pointed out that in most transfer RNA's breakdown of the tertiary structure occurs between 20°C and 40°C, during thermal denaturation; above 40°C secondary structures will melt or

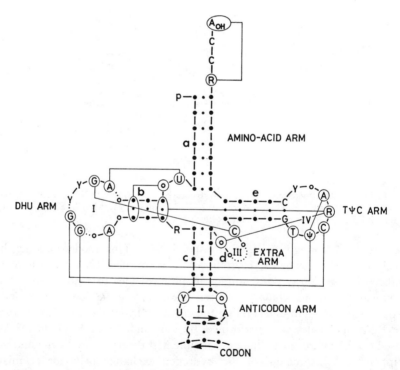

Figure 2-9. A generalized schematic of the clover leaf model of tRNA. I, II, III and IV are the loops or unpaired regions; a, b, c, d and e are the helical or base-paired regions. Solid circles are bases in the helical regions usually paired by hydrogen bonds (indicated by a dot). Open circles are bases usually unpaired. Arrows → indicate 3' to 5' direction; ~ indicates wobble pairing. Nucleotides which interact to form a tertiary structure are circled and joined together. [From Levitt, M., *Nature* **224**, 759 (1969).]

secondary and tertiary structures will melt cooperatively. From their peroxidation and hyperchromicity data Cramer, et al.,[54] interpreted the melting curves of transfer RNA on the structural basis (assumption) that yeast phenylalanine transfer RNA has 24 base pairs and is clover leaf in shape. They suggested that if the second step in the melting curve corresponds to the melting of 24 base pairs, then the 6.1 per cent hyperchromicity of the first step would correspond to eight base pairs or base pair equivalents for the tertiary structure. From the results of the peroxidation experiments they calculated that only four adenosine residues are unpaired in yeast phenylalanine transfer RNA. On the basis of these calculations they postulated a highly ordered structural model for transfer RNA in which the anticodon arm is directed away from the other three arms of the clover leaf, which are in turn folded tightly together.[54]

Other chemical evidence which has been useful in model-building was the demonstration that carbodi-imide reacts with the anticodon and extra loops of the transfer RNA model preferentially,[18] and also that in the TψC loop the ψ is protected from cyanoethylation.[285] On the basis of these and other results from a number of laboratories, Levitt has postulated a very promising model of transfer RNA which fits quite closely the sequences of 14 different transfer RNA's that have been described in the literature.[160] Levitt points out that the probability that the clover leaf arrangement is an artifact of random sequences is less than 1 in 10^{20}. In his model (Figures 2-9 and 2-10) the amino acid arm is stacked on the TψC arm, and the dihydrouracil arm is stacked on the anticodon arm, forming two regular double-helical structures. He considers this arrangement of the arms of the clover leaf, which results in two helical stacks, to be the most stable arrangement since both double-helical structures have an uninterrupted ribose phosphate chain in one of the strands and only one strand has a single nucleotide, the one between the dihydrouracil and anticodon arms, which is stacked but not base-paired.

This model fits the following parameters. The radius of gyration of transfer RNA is approximately 24 Å, which suggests a long, thin molecule.[149] Placing the amino acid-TψC helical stack above the dihydrouracil-anticodon helical stack makes the helical axis parallel. A photoreaction has been shown to occur between 4-tU in position 8 and C in position 13 in several *E. coli* transfer RNA's,[284] and this reaction requires that these bases be brought close to each other in the molecule. This is done by separating the helix axis of the two stacks by only 10 Å. Thus, the diameter of a RNA double helix is about 22 Å. Levitt uses a number of other arguments for the helix interactions. He points out that known transfer RNA sequences can be divided into three classes, based on the structure of the dihydrouracil arm. The model he described is directly applicable to all class 1 sequences which have four base pairs in the dihydrouracil arm, but slight modifications are necessary to accommodate the other classes of sequence. There are 81 hydrogen bonds in his model, which compares very favorably with the data

52 / MOLECULAR STRUCTURE OF THE NUCLEIC ACIDS

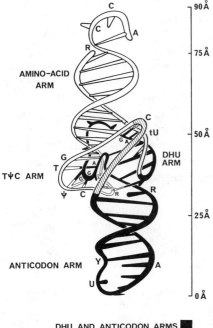

DHU AND ANTICODON ARMS ■
TψC AND AMINO-ACID ARMS □
EXTRA ARM ▨
NUCLEOTIDES 8 AND 9 ▦

Figure 2-10. A diagram of a model of the tertiary structure of tRNA. [From Levitt, M., Nature *224*, 759 (1969).]

of Englander and Englander[73] and optical rotatory dispersion studies of Cantor, Jaskunas and Tinoco.[33]

The long, thin transfer RNA molecule could be oriented to form non-crystalline fibers[64] and could form crystals with hexagonal symmetry, as has been observed for several types of transfer RNA's.[43,108,138] The base 4-tU in position 8 touches cytosine in position 13 and the photoreaction observed by Yaniv, et al.,[284] would be feasible. The number of bases in the loops of the clover leaf and the bases which are exposed agree with the monoperphthalic oxidation data, carbodi-imide reactivity of the anticodon and other loops and the lack of cyanoethylation of ψ in the TψC loop. The long, thin molecule also would allow transfer RNA's to fit side by side on adjacent messenger codons on messenger RNA and to bind in sites proposed for ribosomal structures as discussed in the previous section. It is almost certain that Levitt's model is not the final definitive structure for transfer RNA, but it certainly provides a working model on which other experiments can be designed to alter or modify and ultimately produce a definitive model.

Configuration of 5 s RNA

Speculations concerning the organization of 5 s RNA in the ribosomal structure have already been presented in the section on the configuration of ribosomal RNA. However, 5 s RNA is a separate molecular entity from

CONFIGURATION OF THE NUCLEIC ACIDS / 53

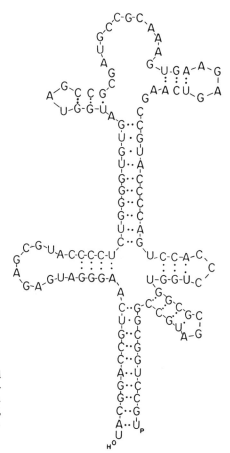

Figure 2-11. A suggested base-paired structure of 5 s RNA. Two dots (··) indicates a standard G–C or A–U base pair. One dot (·) a G·U pair. [From Boedtker, H., and Kelling, D. G., Biochem. Biophys. Res. Commun. 29, 758 (1967).]

ribosomal RNA and can be considered from the point of view of structure and configuration on its own. Since the total sequence of at least two types of 5 s RNA molecules is known,[24,81] this molecule has been studied in a similar way to transfer RNA, although not as extensively. When Brownlee, Sanger and Barrell worked out the primary nucleotide sequence for *E. coli* 5 s ribosomal RNA, they suggested an open ringlike structure for the molecule with three short helical stems[24] (Figure 2-3). By means of hyperchromicity studies on formaldehyde-treated 5 s RNA Boedtker has estimated the minimum base pairing content of 5 s RNA to be 60 per cent and on the basis of this data suggested a structure of 5 s RNA with six looped out regions similar in structure to the clover leaf conformation of transfer RNA[17] (Figure 2-11). He suggests that this conformation may explain why 5 s RNA can compete for one of the transfer RNA binding sites on the ribosome as shown by Comb and Sarkar.[46] Following the determination of the primary structure of 5 s RNA from mammalian KB cells by Forget and

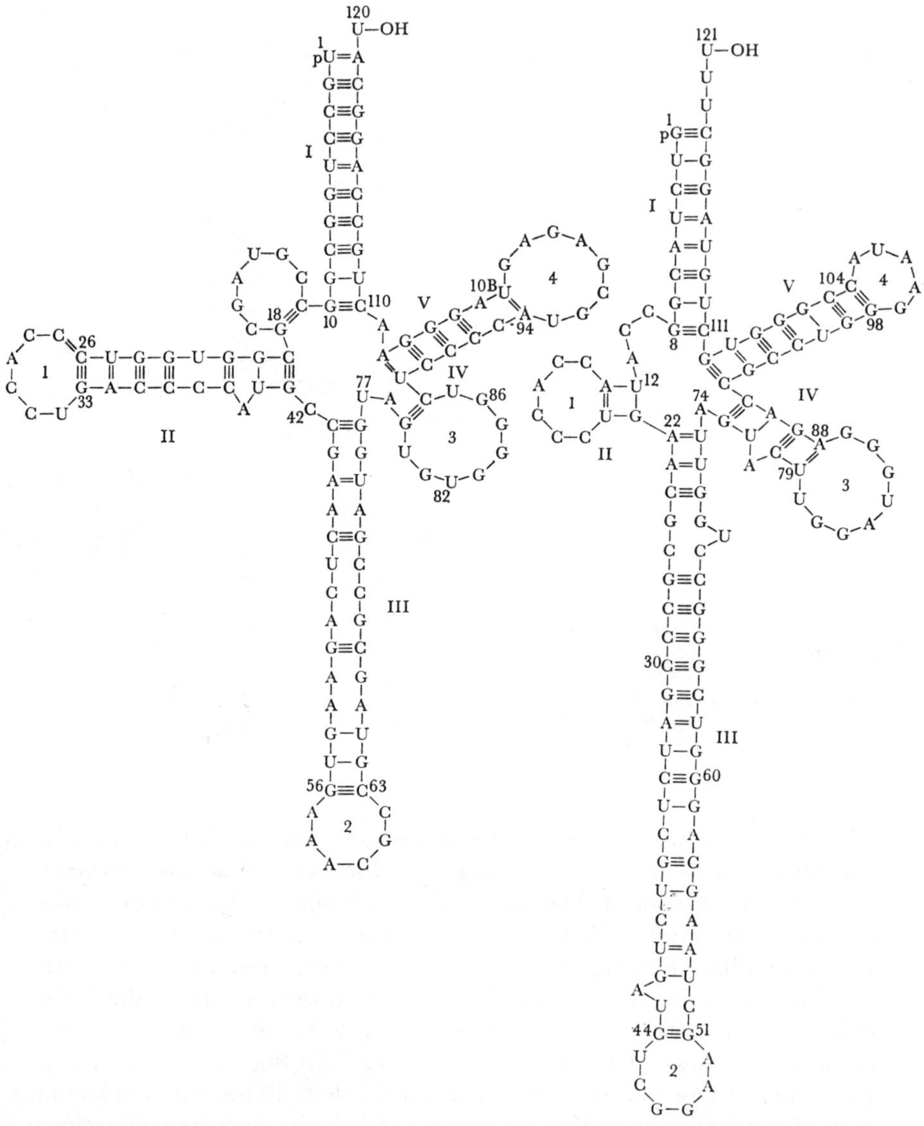

Figure 2-12. Suggested base-paired structures of *E. coli* 5 s RNA and KB cell 5 s RNA. [From Raacke, I. D., Biochem. Biophys. Res. Commun. *31*, 528 (1968).]

Weissman,[81] Raacke suggested another structure for 5 s RNA[201] (Figure 2–12). His structure for *E. coli* has 32 base pairs and an additional 18 non-hydrogen-bonded bases in helical regions, resulting in a total helical content of 68 per cent (Figure 2–12). In his KB model there are 37 base pairs and 12 additional bases in helical regions, resulting in a total helical content of 70 per cent (Figure 2–12), which is in close agreement with the data of Boedtker and Kelling on the base pairing content of 5 s RNA. The models have five looped out regions, imparting five main areas to the structures. Loop 1 in both structures contains the unique sequence CCACC. The deployment of the arms and loops in the model is the same as in the known transfer RNA molecules, which can be folded in various configurations. Loops 2 and 4 in both models are species specific. Loop 3 contains the similar sequence UGGG. Raacke[201] suggests that loop 1 is the site where the 5 s RNA binds to the 50 s ribosome subunit and points out that if this is correct 5 s RNA's from any source could bind to any kind of 50 s ribosome subunit. He also speculates that loop 3 and loop 2 of 5 s RNA extend from the 50 s subunit to the 30 s subunit and that this part of the molecule functions in holding the 30 s and 50 s subunits together. The possibility that 5 s RNA is a spacer or that it holds the two ribosomal subunits together has been discussed in the previous section. Raacke further points out that the common sequence UGGG will pair with the sequence TψCG in loop 4 of the transfer RNA's if the 5 s RNA molecule is rotated by 80 degrees so that the 3' ends of the 5 s RNA and transfer RNA are juxtaposed. He suggests three areas for investigation of the model—the binding of 5 s RNA to 50 s subunits by means of the universal base sequence CCACC, the joining of 30 s to 50 s subunits in a species specific manner by 5 s RNA and the base pairing of transfer RNA and a 5 s RNA in the presence of magnesium by antiparallel base pairing between the common sequences UGGG and TψCG.

A major difficulty in trying to assess the conformational structure of 5 s RNA is that the conformation of the 5 s RNA on the ribosome may be unique and quite different from that in 5 s RNA isolated *in vitro*. Sarkar and Comb, in studying the attachment and release of 5 s RNA, have shown that isolated 5 s RNA will exchange with native 5 s RNA present on 50 s subunits, but the ribosomal subunits with the exchanged 5 s RNA on them are completely inactive in protein synthesis.[218] Although their results are not clear-cut, they do support the concept that the 5 s RNA on the ribosome has a unique conformation.

Configuration of DNA

Knowledge of the structure and configuration of DNA is derived almost entirely from physical studies. Utilizing the X-ray diffraction data of Wilkins and Franklin[85,279] and the chemical data from Chargaff's laboratory,[41] Watson and Crick, in 1952, published the double-stranded helical model of DNA.[270] The model comprises two sugar phosphate esterified chains of

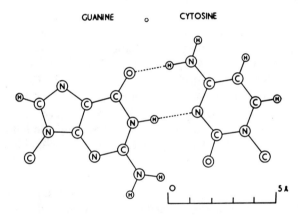

Figure 2-13. Specific hydrogen bonding of the base pairs adenine-thymine and guanine-cytosine, as suggested by Watson and Crick. [From Watson, J. D., and Crick, F. H. C., Cold Spring Harb. Symp. Quant. Biol. *18*, 126 (1953).]

nucleotide residues that are parallel to each other but antiparallel in the direction of orientation of the 3' to 5' phosphodiester linkages. The phosphate residues are oriented towards the outside of the two parallel chains with the nucleic acid bases at right angles to them, projecting inside the space between the two chains so that the bases on the opposite chains are in the same plane and opposite each other. These orientations of base-sugar-phosphate fit Furberg's early experimental X-ray data which defined the structure of cytidylic acid with the base perpendicular to the sugar and the phosphate group in a plane away from the base plane.[87-89] They postulated

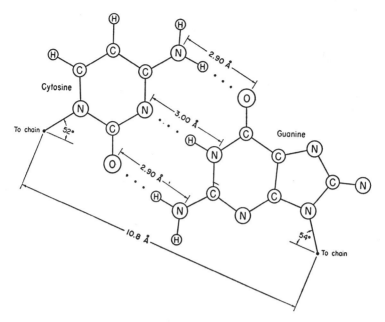

Figure 2-14. Specific hydrogen bonding of the base pair guanine-cytosine according to Pauling and Corey. [From Pauling, L., and Corey, R. B., Arch. Biochem. Biophys. 65, 164 (1956).]

that the pyrimidine base on one strand would be opposite a purine base on the other strand, supporting Chargaff's experimental chemical data of base pairing;[41] thus, adenine would always be opposite thymine, and guanine would always be opposite cytosine. In their original theory Watson and Crick suggested that the bases are joined by hydrogen bonds, with two hydrogen bonds joining each of the two types of base pair (Figure 2-13). This was later modified by Pauling and Corey from other data to two hydrogen bonds between the AT base pair and three hydrogen bonds between the GC base pair[188] (Figure 2-14). The bases are stacked 3.4 Å apart, and the diameter of the two chains is approximately 20 Å. The two parallel chains are coiled in an alpha helical arrangement, forming a shallow groove and a deep groove around the fiber axis of the molecule. One complete turn of the helix comprises 10 base pairs and is 34 Å long (Figure 2-15).

Apart from minor modifications of bond angles and distances this model of the DNA molecule has been the basis for all further work on DNA in this particular area. More recent work on the secondary structure of DNA has centered on the nature of the forces which hold the two strands of the DNA molecule together and interactions between the stacked bases. Hypochromicity studies based on temperature effects, optical rotatory dispersion and circular dichroism measurements on oligonucleotides have shown that stacking forces between the bases play a considerable role in the configuration of the molecule.[10,59,268] Because of the lack of sequence data most of these

Figure 2–15. The helical structure of deoxyribonucleic acid as proposed by Watson and Crick. [From Watson, J. D., and Crick, F. H. C., Nature *171*, 737 (1953).]

studies have been undertaken on synthetic polynucleotides and will be discussed separately in Chapter 3.

The significance of the Watson–Crick model of the secondary structure of DNA is amplified by the fact that all evidence published since the publication of the structure has validated it. Nevertheless, the model has come under scrutiny from a number of workers, and from time to time other models have been suggested.[36,281] The models suggested have not disputed the double-stranded helical structure but have questioned whether there are two strands or four strands associated together in the secondary structure and whether the double-stranded molecule fits in with other interpretations of its function as a genetic template.[47] This second aspect has generated more controversy since the double-stranded helical structure has been adopted as the basis of a number of theories, postulated in particular by Crick and Jacob and Monod,[55,56,130,271] of both protein synthesis and genetic inheritance; if one accepts the basic structure as fact, then all biological evidence should be explainable in terms of this structure.

One of the other alternatives to Watson and Crick's hypothesis of a double-stranded helical model was Cavalieri's suggestion, put forward on the basis of electron microscope data and radioisotope incorporation studies, that DNA is a four-stranded molecule within the chromosomal apparatus.[36] Still another is the reinterpretation of the X-ray diffraction data comparing the 66 per cent with the 93 per cent humidity data by Wu[281] on which he drew the conclusion that on the basis of discrepancies between the two sets of X-ray diffraction data the DNA model could equally well be a complex of two double-stranded molecules. Wu's calculations have recently been re-examined by Wilkins,[278] who maintains that they fit the

CONFIGURATION OF THE NUCLEIC ACIDS / 59

double-stranded interpretation of Watson and Crick far better than Wu's own interpretation of a four-stranded molecule.

Despite the general acceptance of the Watson–Crick model for the secondary structure of DNA, the tertiary structure of DNA, particularly its organization into the chromosomal elements, is still very much a matter of conjecture. Quantitative measurements of the amount of DNA present within cell organelles and viruses compared to the molecular dimensions of the Watson–Crick model show that DNA must be very well organized into these areas within the cell or virus. At the same time it must be able to undertake its biological functions in duplication (replication) and transcription (the synthesis of RNA). The quantitative data has led to speculations on how DNA can fold when still in a double-helical conformation. Numerous suggestions have been made including the suggestion of protein linkers within the DNA molecule;[252] this idea is supported by evidence such as tryptic digestion of high molecular weight DNA isolates and differences in molecular weight estimation by various physical methods, such as sedimentation velocity and light scattering, both before and after trypsin treatment.[252] Bendich has suggested that amino acids of the type which can be phosphorylated, such as serine and threonine, may be incorporated into the sugar phosphate backbone in phosphate ester linkage to provide sufficiently flexible areas to allow bending of the molecule.[12] Other suggestions include single-stranded regions in part of the DNA molecule.[120] In most organisms the DNA has a very high molecular weight and shears even during the most careful isolation; thus, this hypothesis is difficult to test. In some relatively low molecular weight bacteriophage DNA's there are no breaks or single-stranded regions in the molecule.[255] However, the DNA may be nicked; that is, there may be one strand with a scission in the phosphate backbone. This has been shown for a number of bacteriophages.[255] There is no substantial evidence of nicks in bacterial or mammalian DNA mainly because isolation of these DNA's presents difficulties and because of the inadequacies of any controls to show that nicking is not a product of isolation.

The clearest evidence of types of secondary structure have been obtained with the simplest types of DNA. The single-stranded bacteriophages of low molecular weight have circular DNA's[79,255] and it has been shown recently that bacteriophage fd DNA contains only 3', 5' phosphodiester linkages throughout the circle.[219] No other elements, such as phosphorylated amino acids, are present. The development of the monolayer spreading technique for electron microscopic studies on DNA allows excellent studies on the structure of DNA to be carried out.[141] The circularity of the DNA's has been shown by this techique.[86] The single-stranded phages which go through a double-stranded intermediate (replicative form) during their synthesis have also been investigated. It has been shown that the double-stranded replicative forms are of two types, RF I and RF II. RF II has one nick in one of the DNA strands and exists as a circular molecule in an open configuration.[226] The RF 1 form is not nicked, and the two strands of the helix have extra turns, giving the molecule a super-helical configuration so that the molecule under the electron microscope appears as a twisted circle.[140] Specific enzymes are presumed to be present for nicking the DNA, converting the RF I form to the RF II form, and ligases are present which can repair nicks, converting the RF II form into the RF I form.[143] Despite extensive studies which have provided explanation of many facets of the single-stranded phage synthesis,[143,226] the studies are not yet complete and the exact mechanism of phage synthesis is still speculative. However, the presence of the super-helical RF I form shows that the DNA molecule can wrap around itself and has a high degree of flexibility.

Studies on some larger phages, such as the coliphages, have shown that some of these linear phage molecules are comprised of a unique sequence of DNA, while others have reiterated sequences.[255] Also, as mentioned previously, some have been shown to have nicks in the DNA strands. Some coliphages, such as λ phage, upon infection of the bacteria, become circular but are released as linear DNA molecules.[16,287] In bacteriophage λ it has been shown that the two ends of the DNA molecule are single-stranded and complementary to each other.[116,117,280] Thus, the single-stranded portion of one end of the DNA fits into a double-stranded helical base-paired formation with the single-stranded opposite end of the molecule, and in this way a circular DNA molecule can be formed.[209] Linear phage DNA molecules have also been shown to occur in different forms. Some phage DNA's, such as T3, T5, T7 and λ, are non-permuted[1,211] that is, they are unique and each of the DNA molecules has exactly the same sequence in the same order. Other linear DNA molecules, such as P22, T2 and T4, are permuted;[63,254-256] that is, although all the molecules of the same phage have the same overall nucleotide sequence, the order of certain blocks of the sequence can be changed. There is also evidence that the molecules in certain phages, such as T4, T7 and λ, are synthesized in concatenated forms, that is, long repeating chains of the virus in repetitive sequences.[84,274] It is presumed that these concatenates are end to end enchainments of the viruses by their cohesive ends. Mitochondrial DNA's, episomal DNA's from bacteria and some bacterial DNA's have also been shown to be circular, as described previously. All these types are double-stranded, and circularity has been demonstrated by a number of parameters. Usually electron microscopy is the most convincing evidence for circularity, although binding of certain intercalating dyes, particularly ethidium bromide, which result in changes in buoyant density of the closed circular DNA's, has been used extensively.[202] There is no evidence of circularity of the genome in the nuclear DNA of higher organisms.

With regard to the conformation of DNA it is usual to consider only the conformation of isolated DNA in solution. However, DNA occurs in the cell in conjugation with protein. In the bacterial viruses a number of experiments have been performed on the reassociation of the DNA with the protein

CONFIGURATION OF THE NUCLEIC ACIDS / 61

subunits of the virus. Isolated reports of infectious viral particles following reassembly have appeared.[83] Studies on various cation requirements for the reassociation have provided evidence that for reconstitution of spherical viruses, particularly ϕX174, the nucleic acid must have a definite tertiary structure in order to form a complex with the virus protein.

Because of the difficulty of isolation there are virtually no reports on the protein components of mitochondrial DNA. However, the mitochondrial DNA appears to be closely associated in the mitochondria with the mitochondrial membrane[181] and presumably has a protein association in the mitochondrial structure.

In higher organisms DNA occurs in somatic cells in conjugation with a specific class of proteins called histones and in sperm cells with a similar type of protein called protamine. Despite extensive studies over a long period of time the configuration of the DNA with respect to these proteins is still not completely understood. Both the histones and the protamines are extremely basic proteins and are rich in arginine, lysine and glycine. They are associated with the DNA by salt linkages, and the histone can be dissociated from the DNA very easily by high salt concentrations. More recently it has been shown that in the chromosomes of higher animals histones are not the only proteins present in the chromosomal structure, that a class of acidic proteins are also present which are thought to overlay the histones.[113,241] Some of the most recent understanding of the fine structure of the chromosomes has come from the study of RNA synthesis in amphibian oocytes from *Xenopus laevis*, the South African clawed toad, and *Triturus viridescens*, the common spotted newt. The oocyte nucleus is very large, and the contents can be isolated rapidly by manual techniques and observed by both light and electron microscopy. By treatment with low ionic conditions, it is possible to differentiate nucleolar cores and corteces under the electron microscope. When the cores are maximally unwound, they are shown to consist of a thin fiber that is periodically coated with matrix material. Each matrix-covered unit consists of about 100 fibrils, each attached at one end to the core axis, the fibrils forming a gradient of short to long lengths along each matrix unit.[173] The core axis can be removed by deoxyribonuclease, which digests DNA, and the matrix material by ribonuclease, which digests RNA, and protease, which digests protein.[172] It is thought that these matrix fibrils are RNA molecules which are immediately coated with protein following transcription and protein synthesis and that each fibre is the product of a gene in the core axis, the DNA within each matrix-covered segment being a gene coding for the RNA molecule. In the amphibian oocyte the nucleolar DNA is the DNA for ribosomal RNA synthesis.[23,91] There are about 100 ribosomal RNA precursor molecules transcribed simultaneously on each gene in the amphibian oocyte nucleolus. The cross section of the core shows a granular cortex surrounding a compact fibrous core. Results of treatment of the axial fibres with trypsin, a protein-degrading enzyme, suggest that the core axis is a single double-helical DNA molecule coated with protein.[173] These studies provide evidence that in the amphibian oocyte nucleolar region the chromosomal apparatus consists of a very long, repetitious DNA molecule. Hybridization studies with precursor ribosomal RNA indicates that approximately 40 per cent of the nucleolar DNA is inactive and about 60 per cent consists of genes coding for the precursor ribosomal RNA molecules. How these DNA molecules are folded into the chromosome is still a matter of speculation.

REFERENCES

1. Abelson, J. N., and Thomas, C. A., Jr., J. Mol. Biol. *18*, 262 (1966).
2. Apgar, J., Holley, R. W., and Merrill, S. H., J. Biol. Chem. *237*, 796 (1962).
3. Aronson, A. I., J. Mol. Biol. *5*, 453 (1962).
4. Ashwell, M., and Work, T. S., Ann. Rev. Biochem. *39*, 251 (1970).
5. Avers, C. J., Billheimer, F. E., Hoffmann, H.-P., and Pauli, R. M., Proc. Nat. Acad. Sci. (Wash.) *61*, 90 (1968).
6. Baev, A. A., Venkstern, T. V., Mirzabekov, A. D., Krutilina, A. I., Li, L., and Aksel'rod, V. D., Molek. Biol. *1*, 754 (1967).
7. Barnett, W. E., and Brown, D. H., Proc. Nat. Acad. Sci. (Wash.) *57*, 452 (1967).
8. Barnett, W. E., Brown, D. H., and Epler, J. L., Proc. Nat. Acad. Sci. (Wash.) *57*, 1775 (1967).
9. Barrell, B. G., and Sanger, F., FEBS Letters *3*, 275 (1969).
10. Bautz, E. K. F., and Bautz, F. A., Proc. Nat. Acad. Sci. (Wash.) *52*, 1476 (1964).
11. Bazaral, M., and Helinski, D. R., J. Mol. Biol. *36*, 185 (1968).
12. Bendich. A., and Rosenkranz, H. S., Prog. Nucleic Acid Res. Mol. Biol. *1*, 219 (1963).
13. Bernardi, G., Biochim. Biophys. Acta *174*, 435 (1969).
14. Bernardi, G., Carnevali, F., Nicolaieff, A., Piperno, G., and Tecce, G., J. Mol. Biol. *37*, 493 (1968).
15. Bode, H. R., and Morowitz, H. J., Abstr. Biophys. Soc. *9*, 109 (1965).
16. Bode, V. C., and Kaiser, A. D., J. Mol. Biol. *14*, 399 (1965).
17. Boedtker, H., and Kelling, D. G., Biochem. Biophys. Res. Commun. *29*, 758 (1967).
18. Bostoff, S. W., and Ingram, V. M., Science *158*, 666 (1967).
19. Brahic, M., and Fraser, M. J., Biochim. Biophys. Acta *240*, 23 (1971).
20. Brawerman, G., Biochim. Biophys. Acta *72*, 317 (1963).
21. Brimacombe, R. L. C., Griffin, B. E., Haines, J. A., Haslam, W. J., and Reese, C. M., Biochemistry *4*, 2452 (1965).
22. Britten, R. J., and Waring, M. J., Carnegie Inst. Wash. Yearbook *64*, 316 (1965).
23. Brown. D. D., and Dawid, I. B., Science *160*, 272 (1968).
24. Brownlee, G. G., Sanger, F., and Barrell, B. G., J. Mol. Biol. *34*, 379 (1968).
25. Bruskov, V. I., and Kiselev, N. A., J. Mol. Biol. *37*, 367 (1968).
26. Burdon, R. H., Prog. Nucleic Acid Res. Mol. Biol. *11*, 33 (1971).
27. Burdon, R. H., and Clason, A. E., J. Mol. Biol. *39*, 113 (1969).
28. Burton, K., and Petersen, G. B., Biochem. J., *75*, 17 (1960).
29. Burton, K., and Riley, W. T., Biochem. J., *98*, 70 (1966).
30. Cairns, J., J. Mol. Biol. *6*, 208 (1963).
31. Cairns, J., Cold Spring Harb. Symp. Quant. Biol. *28*, 43 (1963).
32. Cairns, J., Stent, G. S. and Watson, J. D., (Eds.) *Phage and the Origins of Molecular Biology*, Cold Spring Harbor Laboratory of Quantitative Biology, Cold Spring Harbor, New York (1966).
33. Cantor, C. R., Jaskunas, S. R., and Tinoco, I., J. Mol. Biol. *20*, 39 (1966).
34. Cape, R. E., and Spencer, J. H., Can. J. Biochem. *46*, 1063 (1968).
35. Cashmore, A. R., and Petersen, G. B., Biochim, Biophys. Acta *174*, 591 (1969).
36. Cavalieri, L. F., and Rosenberg, R. H., Biophys. J., *1*, 323 (1961).
37. Černý, R., Černá, E., and Spencer, J. H., J. Mol. Biol. *46*, 145 (1969).
38. Černý, R., Mushynski, W. E., and Spencer, J. H., Biochim. Biophys. Acta *169*, 439 (1968).

39. Chambers, R. W., Prog. Nucleic Acid Res. Mol. Biol. *11*, 489 (1971).
40. Chao, F. C., and Schachman, H. K., Arch. Biochem. Biophys. *61*, 220 (1956).
41. Chargaff, E., Experientia *6*, 201 (1950).
42. Chargaff, E., Rüst, P., Temperli, A., Morisawa, S. and Danon, A., Biochim. Biophys. Acta *76*, 149 (1963).
43. Clarke, B. F. C., Doctor, B. P., Holmes, K. C., Klug, A., Marcker, K. A., Morris, S. J., and Paradies, H. H., Nature *219*, 1222 (1968).
44. Clayton, D. A., and Vinograd, J., Nature *216*, 652 (1967).
45. Cohen, S. S., Amer. Scientist *58*, 281 (1970).
46. Comb, D. G., and Sarkar, N., J. Mol. Biol. *25*, 317 (1967).
47. Commoner, B., Nature *220*, 334 (1968).
48. Cory, S., Marcker, K. A., Dube, S. K., and Clark, B. F. C., Nature *220*, 1039 (1968).
49. Cory, S., Spahr, P. F., and Adams, J. M., Cold Spring Harb. Symp. Quant. Biol. *35*, 1 (1970).
50. Cox, R. A., Biochem. J. *114*, 743 and 753 (1969).
51. Cox, R. A., and Bonanou, S. A., Biochem, J., *114*, 769 (1969).
52. Cozzarelli, N. R., Kelly, R. B., and Kornberg, A., Proc. Nat. Acad. Sci. (Wash.) *60*, 992 (1968).
53. Cramer, F., Prog. Nucleic Acid Res. Mol. Biol. *11*, 391 (1971).
54. Cramer, F., Doepner, H., Haar, F. v. d., Schlimme, E., and Seidel, H., Proc. Nat. Acad. Sci. (Wash.) *61*, 1384 (1968).
55. Crick, F. H. C., Symp. Soc. Exptl. Biol. *12*, 138 (1958).
56. Crick, F. H. C., J. Mol. Biol. *19*, 548 (1966).
57. Darby, G., Dumas, L. B., and Sinsheimer, R. L., J. Mol. Biol. *52*, 227 (1970).
58. Darby, G. K., Jones, A. S., Tittensor, J. R., and Walker, R. T., Nature *216*, 793 (1967).
59. Devoe, H., and Tinoco, I., Jr., J. Mol. Biol. *4*, 500 (1962).
60. De Wachter, R., and Fiers, W., Nature *221*, 233 (1969).
61. De Wachter, R., Verhassel, J. P., and Fiers, W., Arch. Intern. Physiol. Biochem. *76*, 176 (1968).
62. Dibble, W. E., and Dintzis, H. M., Biochim. Biophys. Acta *37*, 152 (1960).
63. Dintzis, H. M., Borsook, H., and Vinograd, J., in *Microsomal Particles and Protein Synthesis* (R. B. Roberts, Ed.), Pergamon Press, London, p. 95 (1958).
64. Doctor, B. P., Fuller, W., and Webb, N. L. W., Nature *221*, 58 (1969).
65. Doctor, B. P., Loebel, J. E., Sodd, M. A., and Winter, D. B., Science *163*, 693 (1969).
66. Douthit, H. A., and Halvorson, H. O., Science *153*, 182 (1966).
67. Dube, S. K., and Marcker, K. A., European J. Biochem. *8*, 256 (1969).
68. Dube, S. K., Marcker, K. A., Clark, B. F. C., and Cory, S., Nature *218*, 232 (1968).
69. Dube, S. K., Marcker, K. A., Clark, B. F. C., and Cory, S., European J. Biochem. *8*, 244 (1969).
70. Dudock, B. S., and Katz, G., J. Biol. Chem. *244*, 3069 (1969).
71. Dudock, B. S., Katz, G., Taylor, E. K., and Holley, R. W., Proc. Nat. Acad. Sci. (Wash.) *62*, 941 (1969).
72. Ellery, B. W., and Symons, R. H., Nature *210*, 1159 (1966).
73. Englander, S. W., and Englander, J. J., Proc. Nat. Acad. Sci. (Wash.) *53*, 570 (1965).
74. Fellner, P., European J. Biochem. *11*, 12 (1969).
75. Fellner, P., Ehresmann, C., and Ebel, J. P., Cold Spring Harb. Symp. Quant. Biol. *35*, 29 (1970).
76. Fellner, P., Ehresmann, C., and Ebel, J. P., European J. Biochem. *13*, 583 (1970).
77. Fellner, P., Ehresmann, C., and Ebel, J. P., Nature *225*, 26 (1970).
78. Fellner, P., and Sanger, F., Nature *219*, 236 (1968).
79. Fiers, W., and Sinsheimer, R. L., J. Mol. Biol. *5*, 408 (1962).
80. Flamm, W. G., McCallum, M., and Walker, P. M. B., Proc. Nat. Acad. Sci. (Wash.) *57*, 1729 (1967).
81. Forget, B. G., and Weissman, S. M., J. Biol. Chem. *244*, 3148 (1969).
82. Fraenkel-Conrat, H., J. Amer. Chem. Soc. *78*, 882 (1956).
83. Fraenkel-Conrat, H., *The Chemistry and Biology of Viruses*, Academic Press, New York, p. 173 (1969).
84. Frankel, F. R., J. Mol. Biol. *18*, 127 (1966).
85. Franklin, R. E., and Gosling, R. G., Nature *171*, 740 (1953).
86. Freifelder, D., Kleinschmidt, A. K., and Sinsheimer, R. L., Science *146*, 254 (1964).

87. Furberg, S., Nature *164*, 22 (1949).
88. Furberg, S., Acta Cryst. *3*, 325 (1950).
89. Furberg, S., Acta Chem. Scand. *6*, 634 (1952).
90. Galibert, F., Lelong, J. C., Larsen, C. J., and Boiron, M., Biochim. Biophys. Acta *142*, 89 (1967).
91. Gall, J. G., Proc. Nat. Acad. Sci. (Wash.) *60*, 553 (1968).
92. Gal-Or, L., Mellema, J. E., Moudrianakis, E. N., and Beer, M., Biochemistry *6*, 1909 (1967).
93. Gibson, K. D., and Niederman, R. A., Arch. Biochem. Biophys. *141*, 694 (1970).
94. Gierer, A., and Schramm, G., Nature *177*, 702 (1956).
95. Gierer, A., and Schramm, G., Z. Naturforsch. *11b*, 138 (1956).
96. Gilbert, W., J. Mol. Biol. *6*, 374 (1963).
97. Gilham, P. T., J. Amer. Chem. Soc. *84*, 687 (1962).
98. Gilham, P. T., Ann. Rev. Biochem. *39*, 227 (1970).
99. Goebel, W., and Helinski, D. R., Proc. Nat. Acad. Sci. (Wash.) *61*, 1406 (1968).
100. Goldstein, J., and Harewood, K., J. Mol. Biol. *39*, 383 (1969).
101. Gomatos, P. J., and Tamm, I., Proc. Nat. Acad. Sci. (Wash.) *49*, 707 (1963).
102. Goodman, H. M., Abelson, J., Landy, A., Brenner, S., and Smith, J. D., Nature *217*, 1019 (1968).
103. Gould, H. J., and Simpkins, H., Biopolymers *7*, 223 (1969).
104. Granick, S., and Gibor, A., Prog. Nucleic Acid Res. Mol. Biol. *6*, 143 (1967).
105. Habermann, V., Biochim. Biophys. Acta *55*, 999 (1962).
106. Habermann, V., Col. Czech. Chem. Commun. *28*, 510 (1963).
107. Hall, J. B., and Sinsheimer, R. L., J. Mol. Biol. *6*, 115 (1963).
108. Hampel, A., Labanauskas, M., Connors, P. G., Kirgegaard, L., RajBhandary, U. L., Sigler, P., and Bock, R. M., Science *162*, 1384 (1968).
109. Hashimoto, S., Miyazaki, M., and Takemura, S., J. Biochem. (Tokyo) *65*, 659 (1969).
110. Hayashi, Y., Osawa, S., and Miura, K., Biochim. Biophys. Acta *129*, 519 (1966).
111. Hayatsu, H., Takeishi, K., and Ukita, T., Biochim. Biophys. Acta *123*, 445 (1966).
112. Hayatsu, H., and Ukita, T., Biochem. Biophys. Res. Commun. *29*, 556 (1967).
113. Hearst, J. E., and Botchan, M., Ann. Rev. Biochem. *39*, 151 (1970).
114. Henshaw, E. C., J. Mol. Biol. *36*, 401 (1968).
115. Henshaw, E. C., Revel, M., and Hiatt, H. H., J. Mol. Biol. *14*, 241 (1965).
116. Hershey, A. D., and Burgi, E., Proc. Nat. Acad. Sci. (Wash.) *53*, 325 (1965).
117. Hershey, A. D., Burgi, E., and Ingraham, L., Proc. Nat. Acad. Sci. (Wash.) *49*, 748 (1963).
118. Hickson, F. T., Roth, T. F., and Helinski, D. R., Proc. Nat. Acad. Sci. (Wash.) *58*, 1731 (1967).
119. Highton, P. J., Murr, B. L., Shafa, F., and Beer, M., Biochemistry *7*, 825 (1968).
120. Hnilica, L. S., Prog. Nucleic Acid Res. Mol. Biol. *7*, 25 (1967).
121. Hoagland, M. B., Stephenson, M. L., Scott, J. F., Hecht, L. I., and Zamecnik, P. C., J. Biol. Chem. *231*, 241 (1958).
122. Hodnett, J. L., and Busch, H., J. Biol. Chem. *243*, 6334 (1968).
123. Holley, R. W., Apgar, J., Everett, G. A., Madison, J. T., Marquisee, M., Merrill, S. H., Penswick, J. R., and Zamir, A., Science *147*, 1462 (1965).
124. Hradecna, Z., and Szybalski, W., Virology *32*, 633 (1967).
125. Hudson, B., and Vinograd, J., Nature *216*, 647 (1967).
126. Imura, N., Schwam, H., and Chambers, R. W., Proc. Nat. Acad. Sci. (Wash.) *62*, 1203 (1969).
127. Iwamura, T., Prog. Nucleic Acid Res. Mol. Biol. *5*, 133 (1966).
128. Iwanowski, D., St. Petersburg Acad. Imp. Sci. Bull. *35*, 67 (1892), reprinted in English in Phytopathol. Classics *7*, 27 (1942).
129. Jacob, F., Brenner, S., and Cuzin, F., Cold Spring Harb. Symp. Quant. Biol. 28, 329 (1963).
130. Jacob, F., and Monod, J., J. Mol. Biol. *3*, 318 (1961).
131. Jacob, F., and Wollman, E. L., C. R. Acad. Sci. (Paris) *247*, 154 (1958).
132. Jacobson, K. B., Prog. Nucleic Acid Res. Mol. Biol. *11*, 461 (1971).
133. Jaenisch, R., Hofschneider, P. H., and Preuss, A., Biochem, Biophys. Acta *190*, 88 (1969).

134. Jeng, Y., Gelfand, D., Hayashi, M., Shleser, R., and Tessman, E. S., J. Mol. Biol. *49*, 521 (1970).
135. Jeppesen, P. G. N., Nichols, J. L., Sanger, F., and Barrell, B. G., Cold Spring Harb. Symp. Quant. Biol. *35*, 13 (1970).
136. Jones, A. S., Mian, A. M., and Walker, R. T., J. Chem. Soc. (C), 692 (1966).
137. Jones, A. S., Tittensor, J. R., and Walker, R. T., Nature *209*, 298 (1966).
138. Kim, S.-H., and Rich, A., Science *162*, 1381 (1968).
139. Kit, S., J. Mol. Biol. *3*, 711 (1961).
140. Kleinschmidt, A. K., Burton, A., and Sinsheimer, R. L., Science *142*, 961 (1963).
141. Kleinschmidt, A. K., and Zahn, R. K., Z. Naturforsch. *14b*, 770 (1959).
142. Klug, A., and Caspar, D. L. D., Adv. Virus Res. *7*, 225 (1960).
143. Knippers, R., Razin, A., Davis, R., and Sinsheimer, R. L., J. Mol. Biol. *45*, 237 (1969).
144. Kochetkov, N. K., and Budowsky, E. I., Prog. Nucleic Acid Res. Mol. Biol. *9*, 403 (1969).
145. Küntzel, H., and Noll, H., Nature *215*, 1340 (1967).
146. Kurland, C. G., J. Mol. Biol. *2*, 83 (1960).
147. Kurland, C. G., Voynow, P., Hardy, S. J. S., Randall, L., and Lutter, L., Cold Spring Harb. Symp. Quant. Biol. *34*, 17 (1969).
148. Kuwano, M., Kwan, C. N., Apirion, D., and Schlessinger, D., First Lepetit Colloquium on Biology and Medicine: RNA-Polymerase and Transcription (L. Silvestri, Ed.) North-Holland Publishing Company, Amsterdam, p. 222 (1970).
149. Lake, J. A., and Beeman, W. W., J. Mol. Biol. *31*, 115 (1968).
150. Lane, B. G., and Tamaoki, T., Biochim. Biophys. Acta *179*, 332 (1969).
151. Langridge, R. Science *140*, 1000 (1963).
152. Langridge, R., and Gomatos, P. J., Science *141*, 694 (1963).
153. Larsen, C. J., Galibert, F., Lelong, J. C., and Boiron, M., C. R. Acad. Sci. (Paris) *264*, 1523 (1967).
154. Lau, R. Y., and Lane, B. G., Can. J. Biochem. *49*, 431 (1971).
155. Lawley, P. D., and Brookes, P., Biochem. J. *89*, 127 (1963).
156. Lederberg, J., Physiol. Rev. *32*, 403 (1952).
157. Lee, C. S., and Davidson, N., Biochem. Biophys. Res. Commun. *32*, 757 (1968).
158. Lee, C. S., Davidson, N., and Scaletti, J. V., Biochem. Biophys. Res. Commun. *32*, 752 (1968).
159. Lee, J. C., Ho, N. W. Y., and Gilham, P. T., Biochim. Biophys. Acta *95*, 503 (1965).
160. Levitt, M., Nature *224*, 759 (1969).
161. Lewin, S., Arch. Biochem. Biophys. *113*, 584 (1966).
162. MacHattie, L. A., Berns, K. I., and Thomas, C. A., Jr., J. Mol. Biol. *11*, 648 (1965).
163. MacHattie, L. A., Ritchie, D. A., and Thomas, C. A., Jr., J. Mol. Biol. *23*, 355 (1967).
164. Madison, J. T., Ann. Rev. Biochem. *37*, 131 (1968).
165. Madison, J. T., Everett, G. A., and Kung, H., Science *153*, 531 (1966).
166. Matsuda, K., and Siegal, A., Proc. Nat. Acad. Sci. (Wash.) *58*, 673 (1967).
167. McConaughy, B. L., Laird, C. D., and McCarthy, B. J., Biochemistry *8*, 3289 (1969).
168. McLaren, A., and Walker, P. M. B., Nature *221*, 771 (1969).
169. Merril, C. R., Biopolymers *6*, 1727 (1968).
170. Midgley, J. E. M., Biochim. Biophys. Acta *61*, 513 (1962).
171. Midgley, J. E. M., Biochim. Biophys. Acta *108*, 340 (1965).
172. Miller, O. L., Jr., Nat. Cancer Inst. Monogr. *23*, 53 (1966).
173. Miller, O. L., Jr., and Beatty, B. R., Science *164*, 955 (1969).
174. Miura, K.-I., Prog. Nucleic Acid Res. Mol. Biol. *6*, 39 (1967).
175. Mizutani, T., Miyazaki, M., and Takemura, S., J. Biochem (Tokyo) *64*, 839 (1968).
176. Möller, W., Nature *222*, 979 (1969).
177. Moore, R. L., and McCarthy, B. J., J. Bacteriol. *99*, 255 (1969).
178. Moriyama, Y., Hodnett, J. L., Prestayko, A. W., and Busch, H., J. Mol. Biol. *39*, 335 (1969).
179. Mushynski, W. E., and Spencer, J. H., J. Mol. Biol. *52*, 91 (1970).
180. Mushynski, W. E., and Spencer, J. H., J. Mol. Biol. *52*, 107 (1970).
181. Nass, M. M. K., J. Mol. Biol. *42*, 521 (1969).
182. Nemer, M., and Infante, A. A., Science *150*, 217 (1965).
183. Nichols, J. L., and Lane, B. G., Can. J. Biochem. *44*, 1633 (1967).

184. Nomura, M., Mizushima, S., Ozaki, M., Traub, P., and Lowry, C. V., Cold Spring Harb. Symp. Quant. Biol. *34*, 49 (1969).
185. Osawa, S., Biochim. Biophys. Acta *42*, 244 (1960).
186. Osawa, S., Ann. Rev. Biochem. *37*, 109 (1968).
187. Patel, A. B., and Brown, H. D., Nature *214*, 402 (1967).
188. Pauling, L., and Corey, R. B., Arch. Biochem. Biophys. *65*, 164 (1956).
189. Peacock, A. C., and Dingman, C. W., Biochemistry *6*, 1818 (1967).
190. Pene, J. J., Knight, E., Jr., and Darnell, J. E., Jr., J. Mol. Biol. *33*, 609 (1968).
191. Perry, R. P., Proc. Nat. Acad. Sci. (Wash.) *48*, 2179 (1962).
192. Perry, R. P., Nat. Cancer Inst. Monogr. *14*, 73 (1964).
193. Perry, R. P., Prog. Nucleic Acid Res. Mol. Biol. *6*, 219 (1967).
194. Petersen, G. B., and Reeves, J. M., Biochim. Biophys. Acta *129*, 438 (1966).
195. Philippsen, P., Thiebe, R., Wintermeyer, W., and Zachau, H. G., Biochem. Biophys. Res. Commun. *33*, 922 (1968).
196. Pikó, L., Blair, D. G., Tyler, A., and Vinograd, J., Proc. Nat. Acad. Sci. (Wash.) *59*, 838 (1968).
197. Pochon, F., Massoulie, J., and Michelson, A. M., Biochim. Biophys. Acta *119*, 249 (1966).
198. Pogo, A., Pogo, B., Littau, V., Allfrey, V., Mirsky, A. E., and Hamilton, M., Biochim. Biophys. Acta *55*, 849 (1962).
199. Polli, E., Ginelli, E., Bianchi, P., and Corneo, G., J. Mol. Biol. *17*, 305 (1966).
200. Printz, M., and von Hippel, P. H., Proc. Nat. Acad. Sci. (Wash.) *53*, 363 (1965).
201. Raacke, I. D., Biochem. Biophys. Res. Commun. *31*, 528 (1968).
202. Radloff, R., Bauer, W., and Vinograd, J., Proc. Nat. Acad. Sci. (Wash.) *57*, 1514 (1967).
203. RajBhandary, U. L., and Chang, S. H., J. Biol. Chem. *243*, 598 (1968).
204. RajBhandary, U. L., Chang, S. H., Gross H. J., Harada, F., Kimura, F., and Nishimura, S., Fed. Proc. *28*, 409 (1969).
205. RajBhandary, U. L., Chang, S. H., Stuart, A., Faulkner, R. D., Hoskinson, R. M. and Khorana, H. G., Proc. Nat. Acad. Sci. (Wash.) *57*, 751 (1967).
206. Rhaese, H.-J., and Freese, E., Biochim. Biophys. Acta *155*, 476 (1968).
207. Rhaese, H.-J., Freese, E., and Melzer, M. S., Biochim. Biophys. Acta *155*, 491 (1968).
208. Richardson, C. C., Inman, R. B., and Kornberg, A., J. Mol. Biol. *9*, 46 (1964).
209. Ris, H., and Chandler, B. L., Cold Spring Harb. Symp. Quant. Biol. *28*, 1 (1963).
210. Risebrough, R. W., Tissières, A., and Watson, J. D., Proc. Nat. Acad. Sci. (Wash.) *48*, 430 (1962).
211. Ritchie, D. A., Thomas, C. A., Jr., MacHattie, L. A., and Wensink, P. C., J. Mol. Biol. *23*, 365 (1967).
212. Roblin, R., J. Mol. Biol. *36*, 125 (1968).
213. Rosset, R., and Monier, R., Biochim. Biophys. Acta *68*, 653 (1963).
214. Rosset, R., Monier, R., and Julien, J., Bull. Soc. Chim. Biol. *46*, 87 (1964).
215. Rush, M. G., Gordon, C. N., and Warner, R. C., J. Bact. *100*, 803 (1969).
216. Salomon, R., Kaye, A. M., and Herzberg, M., J. Mol. Biol. *43*, 581 (1969).
217. Sanger, F., Brownlee, G. G., and Barrell. B. G., J. Mol. Biol. *13*, 373 (1965).
218. Sarkar, N., and Comb, D. G., J. Mol. Biol. *39*, 31 (1969).
219. Schaller, H., J. Mol. Biol. *44*, 435 (1969).
220. Schlessinger, D., J. Mol. Biol. *7*, 569 (1963).
221. Sedat, J., and Sinsheimer, R. L., J. Mol. Biol. *9*, 489 (1964).
222. Shapiro, H. S., and Chargaff, E., Biochim. Biophys. Acta *23*, 451 (1957).
223. Shapiro, R., Cohen, B. I., and Servis, R. E., Nature *227*, 1047 (1970).
224. Sherrer, K., and Darnell, J. E., Biochem. Biophys. Res. Commun. *7*, 486 (1962).
225. Sherrer, K., Latham, H., and Darnell, J. E., Proc. Nat. Acad. Sci. (Wash.) *49*, 240 (1963).
226. Sinsheimer, R. L., Knippers, R., and Komano, T., Cold Spring Harb. Symp. Quant. Biol. *33*, 443 (1968).
227. Sinsheimer, R. L., Starman, B., Nagler, C., and Guthrie, S., J. Mol. Biol. *4*, 142 (1962).
228. Slayter, H. S., Warner, J. R., Rich. A., and Hall, C. E., J. Mol. Biol. *7*, 652 (1963).
229. Smith, M., Biochem. Biophys. Res. Commun. *10*, 67 (1963).
230. Spahr, P. F., J. Mol. Biol. *4*, 395 (1962).

231. Spahr, P. F., and Tissières, A., J. Mol. Biol. *1*, 237 (1959).
232. Spencer, J. H., Cape, R. E., Marks, A., and Mushynski, W. E., Can. J. Biochem. *47*, 329 (1969).
233. Spirin, A. S., in *Current Topics in Developmental Biology* (A. Monroy and A. A. Moscona, Eds.), Academic Press, New York, Vol. 1, p. 1 (1966).
234. Spirin, A. S., Belitsina, N. V., and Ajtkhozhin, M. A., Zhurnal. Obstchey Biol. *25*, 321 (1964).
235. Srinivasan, P. R., and Borek, E., Prog. Nucleic Acid Res. Mol. Biol. *5*, 157 (1966).
236. Staehelin, M., Experientia *27*, 1 (1971).
237. Staehelin, M., Rogg, H., Baguley, B. C., Ginsberg, T., and Wehrli, W., Nature *219*, 1363 (1968).
238. Stanley, W. M., Science *81*, 644 (1935).
239. Stanley, W. M., Jr., and Bock, R. M., Biochemistry *5*, 1320 (1965).
240. Starr, J. L., and Sells, B. H., Physiol. Rev. *49*, 623 (1969).
241. Stellwagen, R. H., and Cole, R. D., Ann. Rev. Biochem. *38*, 951 (1969).
242. Stutz, E., and Noll, H., Proc. Nat. Acad. Sci. (Wash.) *57*, 774 (1967).
243. Sueoka, N., J. Mol. Biol. *3*, 31 (1961).
244. Sugiura, M., and Takanami, M., Proc. Nat. Acad. Sci. (Wash.) *58*, 1595 (1967).
245. Summers, W. C., Virology *39*, 175 (1969).
246. Summers, W. C., and Szybalski, W., Virology *34*, 9 (1968).
247. Swartz, M. N., Trautner, T. A., and Kornberg, A., J. Biol. Chem. *237*, 1961 (1962).
248. Szekeley, M., and Sanger, F., J. Mol. Biol. *43*, 607 (1969).
249. Takanami, M., J. Mol. Biol. *23*, 135 (1967).
250. Takemura, S., Murakami, M., and Miyazaki, M., J. Biochem. (Tokyo) *65*, 489 (1969).
251. Tamm, C., Hodes, M., and Chargaff, E., J. Biol. Chem. *195*, 49 (1952).
252. Taylor, J. H., in *Molecular Genetics*, Academic Press, New York, Part 1, p. 62 (1963).
253. Tessman, E. S., J. Mol. Biol. *17*, 218 (1966).
254. Thomas, C. A., Jr., and MacHattie, L. A., Proc. Nat. Acad. Sci. (Wash.) *52*, 1297 (1964).
255. Thomas, C. A., Jr., and MacHattie, L. A., Ann. Rev. Biochem. *36*, 485 (1967).
256. Thomas, C. A., Jr., and Rubenstein, I., Biophys. J. *4*, 93 (1964).
257. Tikchonenko, T. I., and Dobrov, E. N., J. Mol. Biol. *42*, 119 (1969).
258. Tissières, A., Watson, J. D., Schlessinger, D., and Hollingsworth, B. R., J. Mol. Biol. *1*, 221 (1959).
259. Tomasz, M., and Chambers, R. W., Biochemistry *5*, 773 (1966).
260. Traut, R. R., Delius, H., Ahmad-Zadeh, C., Bickle, T. A., Pearson, P., and Tissières, A., Cold Spring Harb. Symp. Quant. Biol. *34*, 25 (1969).
261. Ts'o, P. O. P., Bonner, J., and Vinograd, J., J. Biophys. Biochem. Cytol. *2*, 451 (1956).
262. Uziel, M., and Gassen, H. G., Biochemistry *8*, 1643 (1969).
263. Vanyushin, B. F., and Belozersky, A. N., Dokl. Akad. Nauk. SSSR *129*, 944 (1959).
264. Vanyushin, B. F., and Bur'yanov, Y. I., Biokhimiya *34*, 718 (1969).
265. Wagner, E. K., Penman, S., and Ingram, V. M., J. Mol. Biol. *29*, 371 (1967).
266. Waring, M., and Britten, R. J., Science *154*, 791 (1966).
267. Warner, J. R., Rich, A., and Hall, C. E., Science *138*, 1399 (1962).
268. Warshaw, M. M., and Tinoco, I., Jr., J. Mol. Biol. *13*, 54 (1965).
269. Waters, L., and Dure, L., Science *149*, 188 (1965).
270. Watson, J. D., and Crick, F. H. C., Nature *171*, 737 (1953).
271. Watson, J. D., and Crick, F. H. C., Nature *171*, 964 (1953).
272. Weinberg, R. A., and Penman, S., J. Mol. Biol. *38*, 289 (1968).
273. Weiss, B., and Richardson, C. C., J. Mol. Biol. *23*, 405 (1967).
274. Weissbach, A., Bartl, P., and Salzman, L. A., Cold Spring Harb. Symp. Quant. Biol. *33*, 525 (1968).
275. Weissmann, C., and Borst, P., Science *142*, 1188 (1963).
276. Weissmann, C., Borst, P., Burdon, R. H., Billeter, M. A., and Ochoa, S., Proc. Nat. Acad. Sci. (Wash.) *51*, 682 (1964).
277. Weissmann, C., and Ochoa, S., Prog. Nucleic Acid Res. Mol. Biol. *6*, 353 (1967).
278. Wilkins, M. H. F., Proc. Nat. Acad. Sci. (Wash.) *65*, 761 (1970).
279. Wilkins, M. H. F., Stokes, A. R., and Wilson, H. R., Nature *171*, 738 (1953).
280. Wu, R., J. Mol. Biol. *51*, 501 (1970).
281. Wu, T. T. Proc. Nat. Acad. Sci., (Wash.) *63*, 400 (1969).

282. Wyatt, G. R., Biochem. J., *48*, 584 (1951).
283. Yaniv, M., and Barrell, B. G., Nature *222*, 278 (1969).
284. Yaniv, M., Favre, A., and Barrell, B. G., Nature *223*, 1331 (1969).
285. Yoshida, M., and Ukita, T., Biochim. Biophys. Acta *123*, 214 (1966).
286. Yoshida, M., and Ukita, T., Biochim. Biophys. Acta *157*, 455 (1968).
287. Young, E. T., and Sinsheimer, R. L., J. Mol. Biol. *10*, 562 (1964).
288. Young, J. D., Bock, R. M., Nishimura, S., Ishikura, H., Yamada, Y., RajBhandary, U. L., Labanauskas, M., and Connors, P. G., Science *166*, 1527 (1969).
289. Zachau, H. G., Dütting, D., and Feldmann, H., Z. Physiol. Chem. *347*, 212 (1966).

Chapter 3 SYNTHETIC MODEL POLYNUCLEOTIDES

Two developments in nucleic acid chemistry have been basic to studies on the structure and conformation of nucleic acids, the elucidation of the triplet codes for translation of protein from the RNA template and testing of theories of nucleic acid and protein synthesis.

Both developments concern the synthesis of model polynucleotides. The first was the discovery of the enzyme polynucleotide phosphorylase, which catalyzes the polymerization of ribonucleoside diphosphates to form synthetic polyribonucleotides. The second development was the chemical synthesis of oligodeoxynucleotides, which are then used as primers for the synthesis of DNA of known sequence by the enzyme DNA polymerase. The DNA product was in turn used for the synthesis of RNA of known sequence by the enzyme RNA polymerase. Both of these developments surmounted the hurdle of sequence determination of nucleic acids isolated from biological materials. Native nucleic acid sequences are known for a number of transfer RNA's and 5 s RNA's only. The availability of model compounds with known sequences has allowed many specific structural studies to be carried out and numerous reaction mechanisms to be investigated. Before a description of the studies with these model compounds, their synthesis will be discussed.

SYNTHESIS OF MODEL POLYNUCLEOTIDES

Polynucleotide Phosphorylase

Synthetic polyribonucleotides can be synthesized by the enzyme polynucleotide phosphorylase, first discovered in *Azotobacter agilis*[*] by Grunberg-Managao and Ochoa in 1955.[32] The enzyme polymerized ribonucleoside diphosphates according to the equation, $n\text{XDP} \rightleftharpoons (\text{XMP})_n + n\text{P}_i$, where XDP is any

---

[*] *Azotobacter agilis* is also referred to as *Azotobacter vinelandii*.

nucleoside diphosphate, XMP any nucleoside monophosphate and P_i is orthophosphate.[32,33] The reaction is reversible and requires magnesium. The enzyme also requires a RNA primer but this requirement is not absolute. The enzyme can be made primer-dependent by treatment with the proteolytic enzyme trypsin.[61] Polymerization is considered to occur by stepwise addition of the nucleoside diphosphates to the primer.[115,116]

The enzyme is widely distributed among both aerobic and anaerobic bacteria and is usually prepared from *Micrococcus lysodeikticus* or *E. coli*. The activity of the enzyme varies with the stage of growth of the microorganisms and is highest at the beginning of the logarithmic growth phase.[31] Considered for a long time a bacterial enzyme, its activity has been occasionally reported in animal tissues.[18,39] Recently the enzyme has been purified from guinea pig liver.[22,110] The enzyme is unstable in crude extract form, but stable after partial purification, and is capable of catalyzing the synthesis of polyribonucleotides when only partially pure, provided the nucleoside diphosphate substrate is in excess and phosphatase activity is minimal.[31,33]

Preparation and Assay

The usual method of preparation of polynucleotide phosphorylase involves lysis of the bacterial cells by lysozyme treatment (*M. lysodeikticus*) or sonication (*E. coli*), fractionation by precipitation with ammonium sulfate, DEAE cellulose anion exchange chromatography and Sephadex gel filtration.[31,62] The enzyme can be stored and is stable when frozen for many months after purification to the ammonium sulfate fractionation step. The enzyme is available commercially, and the commerical preparation can be further purified by chromatography on Sephadex G200 to increase the specific activity of the enzyme.[61]

The enzyme can be assayed by a number of methods, the majority of which depend upon measurement of one of the components of the polymerization reaction—either disappearance of substrate (nucleoside diphosphate), appearance of inorganic phosphate or appearance of polymer. The enzyme can also be assayed by measurement of arsenolysis of nucleoside diphosphates since arsenate can replace orthophosphate as a substrate.[113] The most useful and frequently used assays for the enzyme are incorporation of radioactively labeled nucleoside diphosphate substrate into an acidic insoluble precipitate (the polymer)[73] or measurement of the phosphorolysis of poly A, which is followed by absorption on charcoal of ^{32}P-labeled nucleotides[73,114] or by disappearance of ^{32}P orthophosphate from the incubation mixture.[33]

Polymerization Reaction

Although large quantities of polynucleotide can be prepared relatively easily from commercially available nucleoside diphosphates with the commercially available enzyme, there are a large number of factors that can affect the activity of the enzyme and must be taken into consideration during polymerization reactions. There is an absolute requirement for magnesium in all reactions catalyzed by the enzyme, and the pH optimum can vary with the source of the enzyme and with the salt and magnesium concentrations in the reaction medium.[4] The pH optimum for polymerization with the *M. lysodeikticus* enzyme is between 9 and 10[118] and for the *A. agilis* enzyme the range is between pH 7.5 and 9 in the presence of 0.01 M ADP and 5 millimolar magnesium chloride.[31] The concentration of nucleoside diphosphate required to saturate the enzyme is very high. At pH 8.1 saturating conditions for adenosine-5'-diphosphate and inosine-5'-diphosphate are of the order of 0.1 and 0.05 M, respectively, for both the *E. coli* and *A. agilis* enzymes.[33,73] The relationship between the concentrations of nucleotide diphosphate and magnesium is complex. A typical polymerization reaction mixture contains 200 millimoles of tris(hydroxymethyl)aminomethane buffer at pH 8.1, 3 millimoles of magnesium chloride, 1.2 millimoles of ethylene diaminetetraacetic acid, plus enzyme and substrate, plus primer if required. The synthesis reaction is followed closely until synthesis of the polymer is complete, at which time the polymer is immediately isolated from solution by quantitative precipitation in the cold with acid, ethanol, high salt or streptomycin. The polymer precipitate is removed by centrifugation, deproteinized by chloroform treatment, dried *in vacuo* and stored at minus 20°C. It is important to precipitate the product immediately upon completion of the synthetic reaction. Any waiting period after completion of synthesis results in degradation of the synthesized polymer by the reverse reaction of the enzyme.

Substrate Specificity

The enzyme shows specificity for the sugar and diphosphate moieties of the ribonucleoside-5'-diphosphates, but shows little specificity for the bases. Homopolymers (that is, polymers containing only one type of nucleic acid base) are readily synthesized, and the enzyme polymerizes a number of nucleoside analogues, such as inosine-5'-diphosphate,[32,73] N-methyluridine-5'-diphosphate,[123] 5-ethyluridine-5'-diphosphate,[122] 2-thiouridine-5'-diphosphate,[68] 5-fluorouridine-5'-diphosphate, 5-bromouridine-5'-diphosphate, 5-chlorouridine-5'-diphosphate and 5-iodouridine-5'-diphosphate,[75,80] ribosylthymine-5'-diphosphate,[29] 8-azaguanine-5'-diphosphate,[69] xanthosine-5'-diphosphate,[81] N-6 hydroxyethyladenosine-5'-diphosphate,[81] 5-methylcytosine-5'-diphosphate,[125] 2'-O-methylcytosine-5'-diphosphate,[52] 2'-O-methyladenine-5'-diphosphate[103] and 2'-O-methyluridine-5'-diphosphate.[139] The ability to polymerize nucleoside analogues has greatly increased the importance of the enzyme in studies of the structure of nucleic acids.

In addition to forming homopolymers using only one nucleoside diphosphate as substrate, the enzyme will polymerize two, three or four nucleoside diphosphates according to the proportions of the individual nucleoside diphosphates in the reaction mixture. This aspect of its action has been of importance in determination of the coding triplets in protein synthesis. However, the sequence of bases in this type of polymer is random and shows minor variations.

The enzyme also shows specificity towards the 5' phosphorylated nucleoside. Nucleoside diphosphates also esterified at the 3' hydroxyl position with phosphate such as ppUp or ppAp are not substrates but interact with the enzyme since they inhibit the exchange reaction with other diphosphates.[30,31]

Secondary Reactions

The enzyme, as well as catalyzing polymerization of nucleotides, also catalyzes phosphorolysis of polynucleotides, which may be represented by the following reaction:

$$pApApApU \rightarrow pApApA + UDP \rightarrow pApA + ADP$$

Indications are that the phosphorolysis reaction proceeds through a stepwise cleavage and represents a true reversal of polymerization. Phosphorolysis stops at the dinucleotide level and dinucleotides are not phosphorylized.[116]

The enzyme also catalyzes an exchange reaction between orthophosphate and nucleoside diphosphates. This reaction is very dependent on the molar ratio of nucleoside diphosphate to orthophosphate, this ratio in turn being dependent on the magnesium chloride concentration.[118]

The *M. lysodeikticus* enzyme also catalyzes the arsenolysis of polynucleotides, as described earlier, whereby arsenate replaces orthophosphate in the reaction catalyzing the arsenolysis of poly(A), poly(U), poly(C) and RNA to the corresponding 5' mononucleotides.[36,118] It has been suggested that this reaction occurs through the formation of an unstable nucleoside monophosphate-monoarsenate intermediate.[31] There are indications that poly(G) is not arsenolyzed. Arsenate serves as a catalyst in the reaction and the products are nucleoside 5' monophosphates. The replacement of Pi by arsenate in the reaction results in the arsenolysis of nucleoside diphosphates to nucleoside monophosphates.[112]

Reports have also appeared that the enzyme also catalyzes transnucleotidation reactions,[37,117] which may be represented by the following reaction:

$$pApApA + pApApA \rightarrow pApApApA + pApA$$

These secondary reactions of the enzyme are important only in that they can be used as measures of enzyme activity.

Synthesis by Other Enzymes

Model polynucleotides have also been synthesized using other enzymes. RNA polymerase can be used to synthesize polyribonucleotides of known sequence by the use of small chemically synthesized oligodeoxynucleotides as primers.[12] Reversal of the normal degradative reaction of ribonuclease has been used to synthesize a variety of di- and tri-ribonucleotides.[5,38]

Under suitable conditions and without a primer DNA polymerase can be used for synthesis of poly(dA-dT),[94] an alternating co-polymer, and poly(dG), poly(dC) homopolymers.[106] The enzyme can also be used for polymerization of a variety of deoxyanalogues, and poly(dA) has been prepared using a primer.[8] Deoxypolymers have also been synthesized with E. coli DNA polymerase using a ribopolymer template.[65,66] Some of these methods have been superseded with the discovery in calf thymus tissue of the enzyme DNA terminal transferase.[53,63,135] The enzyme displays a substrate preference for adenosine-5'-triphosphate and has been used extensively for preparation of polydeoxyadenylic acid molecules synthesized onto the 3'-OH end of oligodeoxynucleotides of specific sequence or composition.[8]

Chemical Synthesis of Polynucleotides

Attempts to synthesize polyribonucleotides by chemical methods have not been as successful as those for polydeoxyribonucleotides. Michelson has synthesized oligoribonucleotides from nucleoside 2', 3' cyclic phosphate compounds in the presence of tetraphenylpyrophosphate and an anhydrous medium. The products of the reaction are equimolar ratios of the 2', 5' and 3', 5' phosphodiester linkages, and distribution of the two types of linkage appears to be random throughout the polymer.[78,79] The terminal nucleotide is a 2', 3' cyclic phosphate. Smith and Khorana have reported the synthesis of urydilyl-uridine from uridine-3', 5'-phosphate,[119] and investigators in Khorana's laboratory have synthesized all sixty-four possible nucleotide triplets as part of their study of the genetic code.[54]

The success with the chemical synthesis of deoxyribonucleotides is partly due to the absence of the C-2' hydroxyl group in deoxyribose. As noted in Chapter 1, the absence of this group eliminates the possibility of cyclization reactions of the type common in ribonucleotides, and in chemical synthetic methods reduces the number of groups needing protection by one. Following the pioneering work of Michelson and Todd in chemical syntheses of deoxythymidine dinucleotides,[83] Khorana developed a number of procedures for synthesizing deoxyribonucleotide polymers of specific base sequence. These methods have been used to synthesize deoxyoligonucleotides corresponding to parts of the gene sequence for yeast alanine transfer RNA, deduced from the sequence of the transfer RNA. By means of these synthesized deoxyoligonucleotides and the enzyme DNA polymerase, relatively large amounts of polymer material were prepared. Combination of this and other methodology, described later, has resulted in the total synthesis of the gene for yeast alanine transfer RNA.[2] The availability of deoxyoligonucleotides of known sequence has also been used extensively in determination of the amino acid triplet code (genetic code) by using them as templates for the enzymatic synthesis of ribopolynucleotides of known sequence.[56] Polymers of defined sequence are currently used in the study of nucleic acid structure and enzyme specificity.

Condensation Reactions

Development of synthetic procedures for deoxyoligonucleotides required satisfactory methods for the activation of the phosphomonoester group in a mononucleotide so as to cause the phosphorylation of the hydroxyl group of the condensing nucleoside or nucleotide, the design of suitable protecting groups for the various functional groups of nucleotides such as primary and secondary hydroxyl groups, amino groups and phosphomonoester groups, the development of methods for the polymerization of mononucleotides and of methods for the separation and characterization of the resulting polynucleotides, and finally the step-wise synthesis of the polydeoxynucleotides containing specific sequences.

In early studies Khorana used *para*-toluenesulfonyl chloride(I) as the

I
para-Toluenesulfonyl chloride

phosphorylating agent, but subsequent studies showed that dicyclohexylcarbodi-imide (DCC) (II) was superior to the *para*-toluene derivative.[58]

II
Dicyclohexylcarbodi-imide

A detailed study of the properties and mechanism of the reaction was carried out by Gilham and Khorana.[26]

The protecting groups used for the various nucleotides depend upon the parent nucleotide and the groups to be protected. For example, the amino group of adenine can be protected by a N-benzoyl group, and the 3′-OH group of the deoxyribose can be protected by an acetyl group by acetylation of N-benzoyl deoxyadenosine in pyridine, giving the compound N-benzoyl-3'-O-acetyldeoxyadenosine-5'-phosphate (III).[95] The 5' hydroxyl group of

III
N-Benzoyl-3'-O-acetyldeoxyadenosine-5'-phosphate

deoxyribose can be protected by *p*-methoxytrityl groups.[26,27] All the synthetic methods developed in Khorana's laboratory require anhydrous conditions, and anhydrous pyridine has proved to be an effective solvent system. Thus, for condensation of an adenine nucleotide with a thymine nucleotide[132] N-benzoyl-3'-O-acetyldeoxyadenosine-5'-phosphate (III) is condensed with 5'-O-di-*p*-methoxytritylthymidine (Figure 3-1, IV) in the presence of dicyclohexylcarbodi-imide (II) in anhydrous pyridine. Brief alkaline treatment of the product causes selective removal of the acetyl group at the 3' position of the adenine moiety, and the final product is 5'-O-dimethoxytritylthymidylyl-(3' → 5')-N-benzoyldeoxyadenosine (Figure 3-1, V). This compound can in turn be condensed with other suitably protected compounds. One such condensation reaction occurs with the 3'-O-acetylated trinucleotide pTpTpT,[95] resulting in the synthesis of a pentanucleotide (Figure 3-2). Since the trinucleotide is insoluble in dry pyridine, the solvent system used was a mixture of anhydrous pyridine and dimethylformamide. The protecting groups were removed by treatment with ammonium hydroxide and the products separated by chromatography on DEAE cellulose. The yield of the pentanucleotide product, dimethoxytrityl pTpApTpTpT (Figure 3-2, VI), was 12 per cent. Partial proof of structure of the product was made by brief acid hydrolysis with HCl at 25°C, which caused selective removal of the adenine which was characterized by chromatography and spectrophotometric analysis.[132]

Another interesting synthetic deoxynucleotide prepared in Khorana's laboratory is the deoxyadenine-thymine alternating co-polymer.[132] In this synthesis the thymine nucleoside 5'-phosphate was protected by a β-cyanoethyl group by reaction of thymidine-5'-phosphate in the presence of DCC

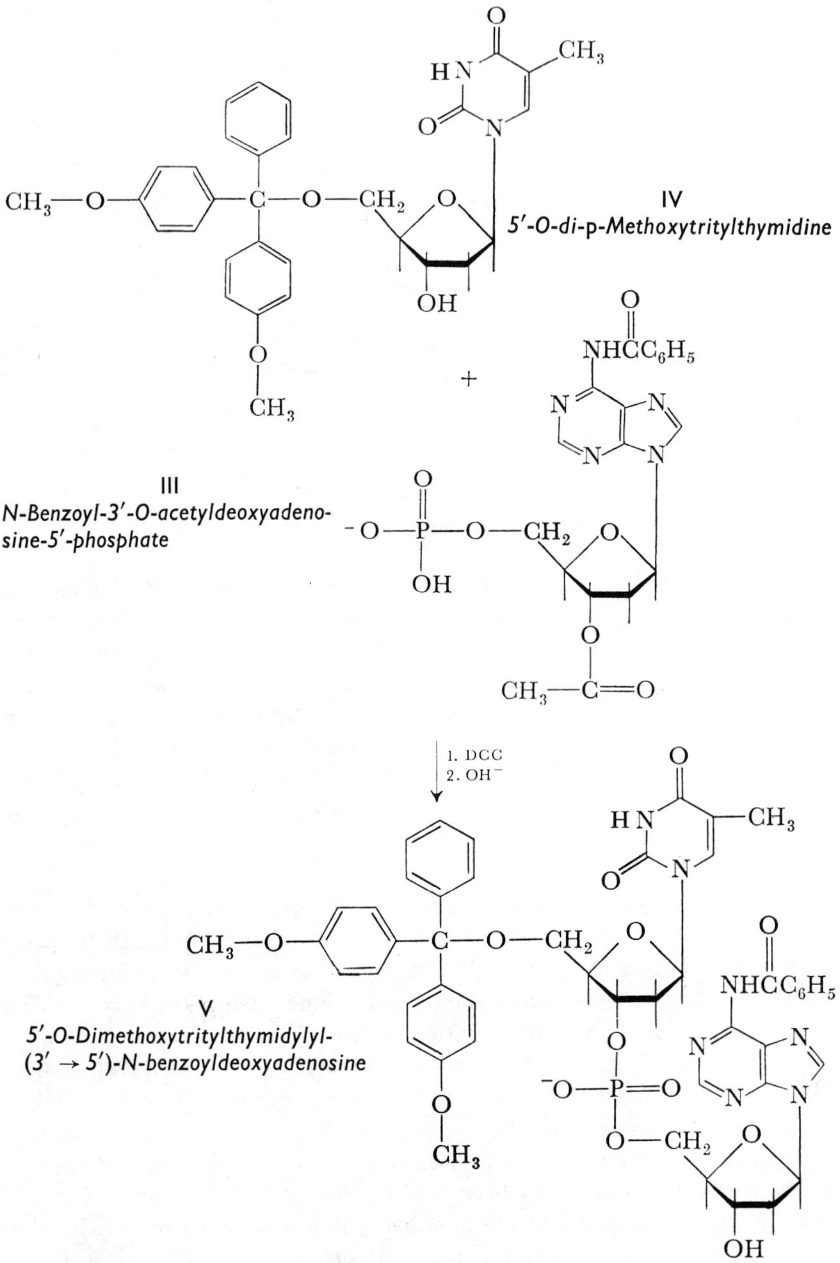

Figure 3-1. [From Weimann, G., Schaller, H., and Khorana, H. G., J. Amer. Chem. Soc. *85*, 3835 (1963).]

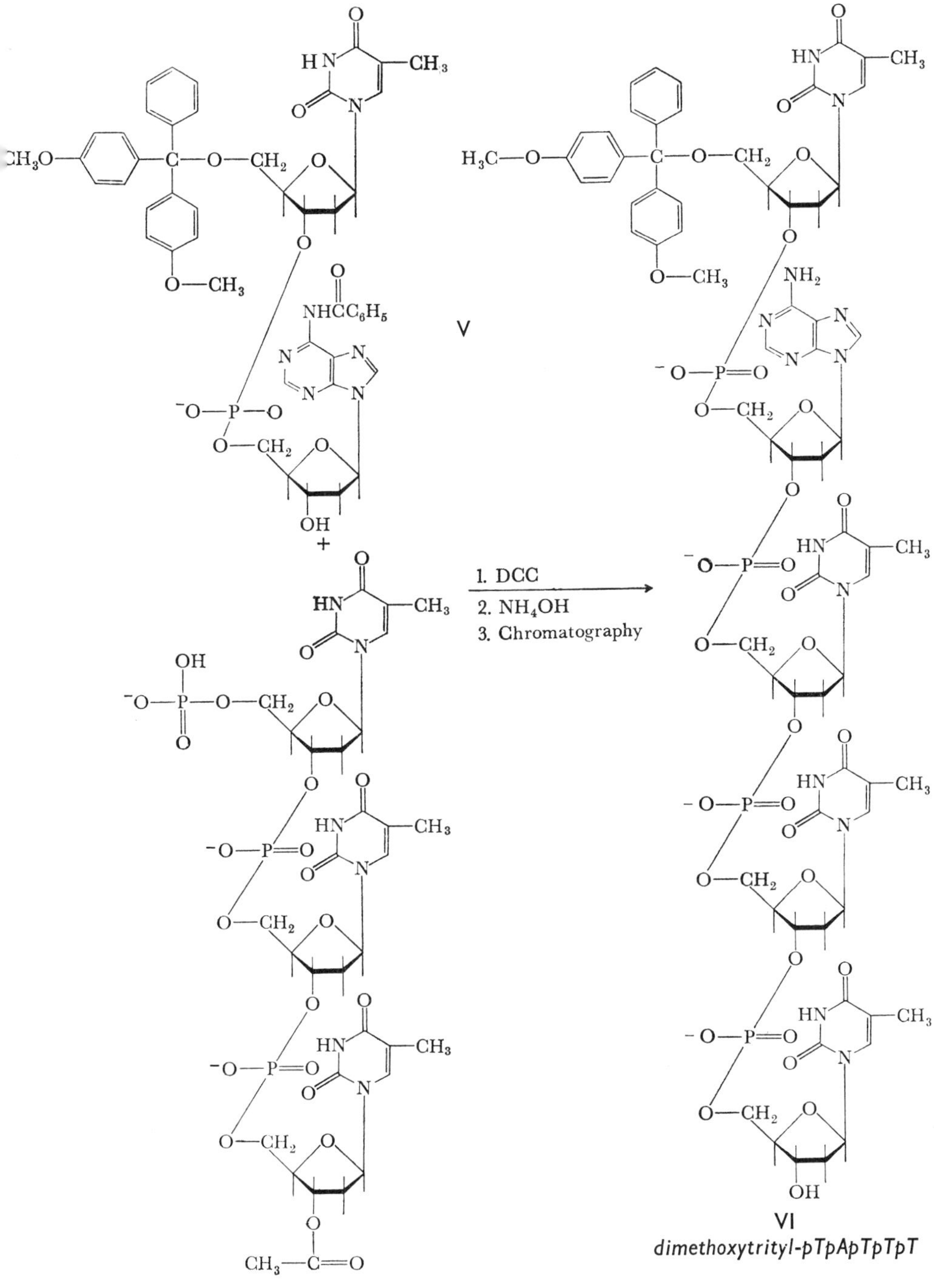

SYNTHESIS OF MODEL POLYNUCLEOTIDES / 77

3'-O-acetyl-pTpTpT

1. DCC
2. NH₄OH
3. Chromatography

VI
dimethoxytrityl-pTpApTpTpT

Figure 3–2.

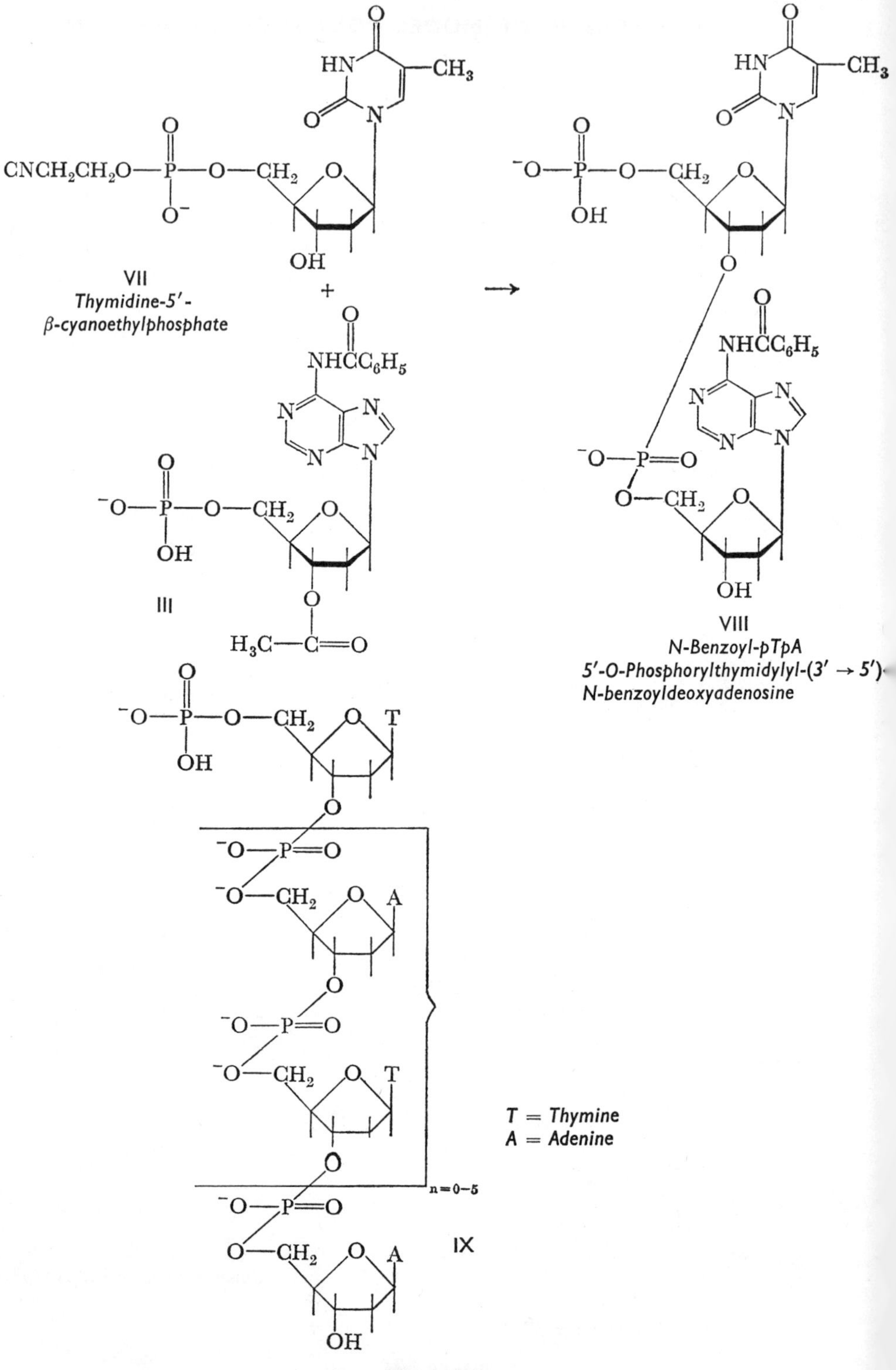

VII
Thymidine-5'-β-cyanoethylphosphate

III

VIII
N-Benzoyl-pTpA
5'-O-Phosphorylthymidylyl-(3' → 5')-N-benzoyldeoxyadenosine

T = Thymine
A = Adenine

n = 0–5

IX

Figure 3–3.

SYNTHESIS OF MODEL POLYNUCLEOTIDES / 79

and an excess of hydroacrylonitrile and pyridine. Condensation of thymidine-5'-β-cyanoethyl phosphate (Figure 3-3, VII) with N-benzoyl-3'-O-acetyldeoxyadenosine-5'-phosphate (III) in DCC followed by careful alkaline treatment to remove the 3'-O-acetyl and the β-cyanoethyl group resulted in the dinucleotide pTpA with the amino group of adenine still protected by the N-benzoyl group (Figure 3-3, VIII). Yields in this type of reaction vary between 30 and 40 per cent. Polymerization of the protected dinucleotide 5'-O-phosphorylthymidylyl-(3' → 5')-N-benzoyl deoxyadenosine (Figure 3-3, VIII) was then possible using the dicyclohexylcarbodiimide condensation method,[60] similar to the example in Figure 3-2, to give linear polynucleotides of the general structure shown in Figure 3-3, IX. Following condensation the reaction mixture was treated with ammonium hydroxide to remove the N-benzoyl protecting groups, and the final polynucleotide products were separated by chromatography on a DEAE cellulose column according to chain length.[126] Compounds up to the dodecanucleotide level have been synthesized by this method; however, yields do not exceed the 15 to 25 per cent range.

Step-wise Synthesis

To overcome the problem of low yields in the condensation reactions Khorana has also developed the use of step-wise synthetic methods.[27, 50] Successful results have been obtained with thymine nucleotides starting with trityl thymine, the trityl group protecting the 5' hydroxy group of the sugar moiety. Condensation of the 5'-O-tritylthymidylic acid with an excess of 3'-O-acetylthymidylic acid in the presence of dicyclohexylcarbodi-imide results in the synthesis of a trityl dinucleotide,[58] which, following further reaction with an excess of 3'-O-acetylthymidylic acid, yields a trityl trinucleotide, and so on[50] (Figure 3-4). Yields were in the range of 70 to 90

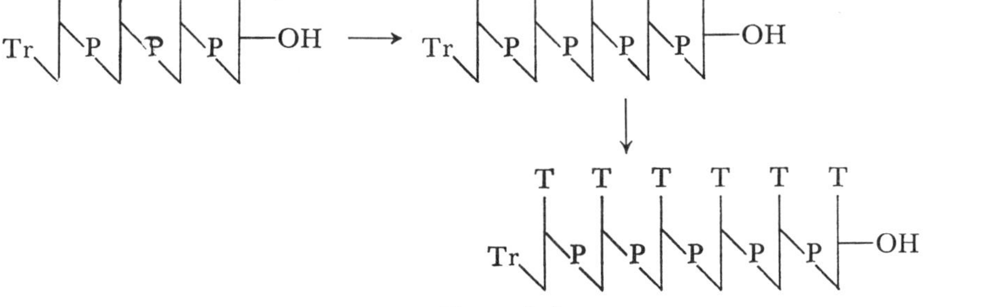

Figure 3-4.

per cent and synthesis up to the hexanucleotide level in DCC is in the 60 per cent range. When DCC was replaced as the polymerizing agent with mesitylene sulfonyl chloride[49] the yield of the hexanucleotide was about 66 per cent. However, all these reactions result in the production of side products, which at the hexanucleotide level of synthesis were present in substantial amounts.[50] Two of the side products were 5'-C-pyridinium-TpT and 5'-C-pyridinium-T (X).

X
5'-C-Pyridinium thymidine

Step-wise condensation has also been used for the synthesis of a polydeoxynucleotide containing the repeating trinucleotide sequence thymidylyl-(3' → 5')-thymidylyl-(3'→ 5')-deoxycytidine (d-TpTpCpTpTpCpTpTpCp-TpTpC).[51] In these series of condensations the 5' hydroxyl group of thymidylic acid was protected with a trityl group and the amino group of cytidylic acid protected with a N-anisoyl group formed by reaction of cytidylic acid with anisoyl chloride (Figure 3–5 XI).[107] The condensing agent used was mesitylene sulfonyl chloride rather than dicyclohexyl-carbodi-imide.[49,50] In the first series of reactions a molecule of 5'-O-tritylthymidine was condensed with a molecule of 3'-O-acetylthymidylic acid and the product treated briefly with alkali (Figure 3–5). The product was then condensed with N-anisoyl-3'-O-acetyldeoxycytidine-5'-phosphate (Figure 3–5, XI). Treatment of the product of this reaction with ammonium hydroxide and then acid resulted in the product trityl-TpTpC. Yield of product was high, ranging between 70 and 80 per cent. Side reactions were reduced and yields increased by use of an increasing excess of the protected mononucleotide, which is readily available in quantity. For longer oligonucleotides the reactions were repeated on the product of each end reaction; thus, trityl-TpTpC was reacted with 3'-O-acetylthymidylic acid, the tetranucleotide trityl-TpTpCpT isolated, treated with 3'-O-acetyl-thymidylic acid again to give the pentanucleotide trityl-TpTpCpTpTpT, which was treated with N-anisoyl-3'-O-acetylcytidylic acid to give the hexanucleotide trityl-TpTpCpTpTpC. The yield at the decanucleotide level was 70 per cent and at the dodecanucleotide level 56 per cent.

SYNTHESIS OF MODEL POLYNUCLEOTIDES / 81

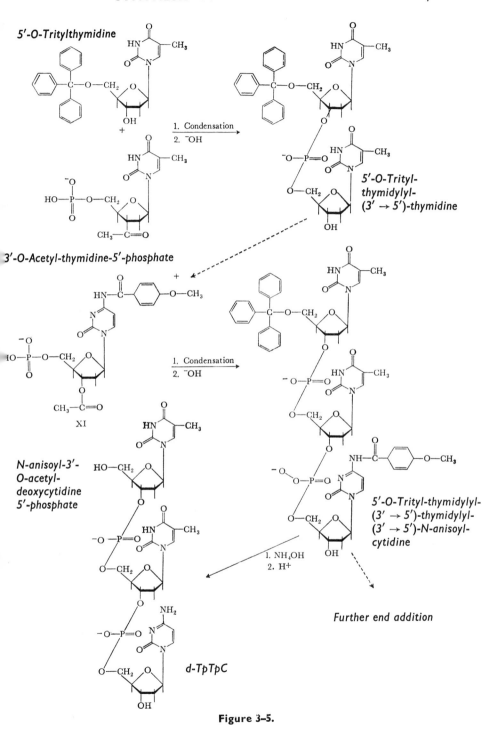

Figure 3–5.

Khorana has used these methods of step-wise condensation for the synthesis of a variety of polydeoxynucleotides with repeating trinucleotide and tetranucleotide sequences which were used to establish the genetic code.[54,56] A list of those used is given in Table 3–1.

Table 3–1. Synthetic Deoxyribopolynucleotides with Repeating Sequences

Repeating Trinucleotide Sequences

$\{d(TTC)_4\}$ $\{d(TAC)_{4-6}\}$
$\{d(AAG)_4\}$ $\{d(TAG)_{4-6}\}$
$\{d(TTG)_{4-6}\}$ $\{d(ATC)_{3-5}\}$
$\{d(CAA)_{4-6}\}$ $\{d(ATG)_{3-5}\}$
$\{d(CCT)_{3-5}\}$ $\{d(CCA)_{3-5}\}$
$\{d(GGA)_{3-5}\}$ $\{d(GGT)_{3-5}\}$
$\{d(CGA)_{3-5}\}$
$\{d(CGT)_{3-5}\}$

Repeating Tetranucleotide Sequences

$\{d(TTTC)_3\}$
$\{d(AAAG)_{3-4}\}$
$\{d(TCTA)_3\}$
$\{d(TAGA)_2\}$
$\{d(TTAC)_4\}$
$\{d(GTAA)_2\}$

* After Khorana, H. G., Büchi, H., Ghosh, H., Gupta, N., Jacob, T. M., Kössel, H., Morgan, R., Narang, S. A., Ohtsuka, E., and Wells, R. D., Cold Spring Harb. Symp. Quant. Biol. *31*, 39 (1966).

Synthesis on a Polymeric Support

Other studies in Khorana's laboratory developed following the very successful experiments by Merrifield on the synthesis of peptides on an insoluble polymeric support.[77] The reason for investigating oligonucleotide synthesis on a polymer support is that, although satisfactory syntheses have been carried out by step-wise condensation reactions, there are a number of problems with regard to side reactions and yield of product which possibly could be overcome by solid phase methods. The main drawback to the step-wise condensation reactions is that each condensation to form internucleotide bonds requires separation of the synthesized material by time-consuming procedures which involve many column chromatographic operations.

Khorana chose a soluble polymer as a support for oligonucleotide synthesis similar to the soluble polymer support used by Shemyakin and co-workers for polypeptide synthesis.[111] A similar approach has been reported by Cramer and co-workers.[16] The polymer chosen was polystyrene having *p*-methoxytrityl groups as a part of its structure.[34] The *p*-methoxytrityl groups of the polystyrene are used for covalent linkage of the first nucleoside or nucleotide to the polymer support in reactions analogous to those worked out for step-wise condensation reactions. A second consideration is that the

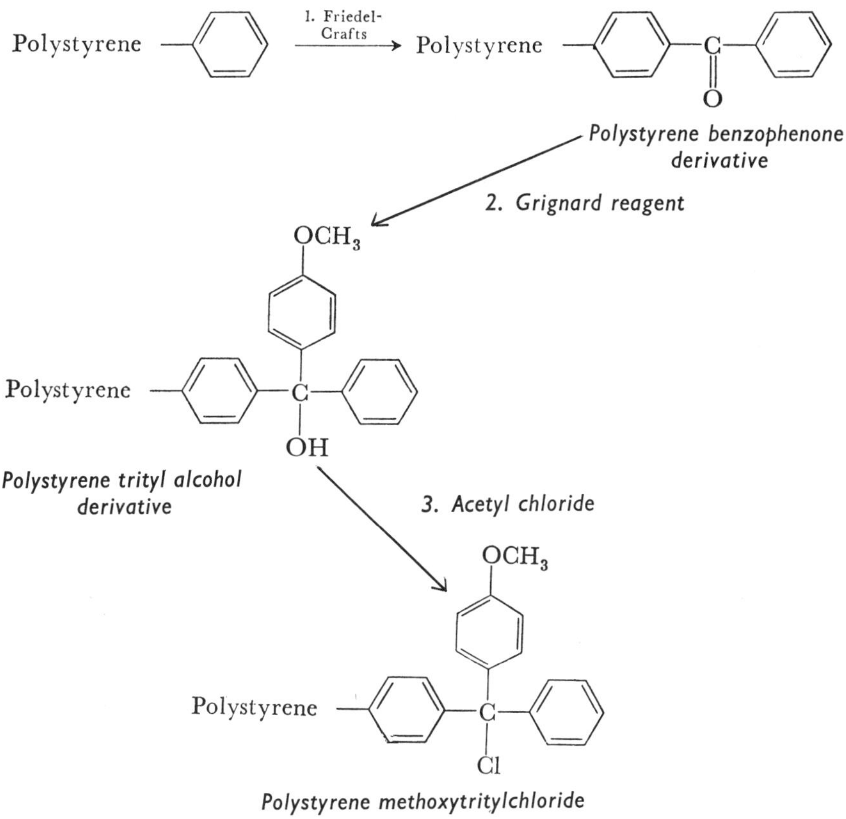

SYNTHESIS OF MODEL POLYNUCLEOTIDES / 83

Figure 3–6.

methoxytrityl derivative of polystyrene is soluble in anhydrous pyridine, which had previously been used as the reaction medium in step-wise polynucleotide synthesis experiments. Also, the extent of introduction of the phenyl groups in polystyrene to form the benzophenone and the subsequent formation of the methoxytrityl derivatives could be controlled by varying the reaction conditions, allowing flexibility with regard to the type of polymer support used in the final reactions.[34]

For preparation of the polymer support polystyrene was subjected to a Friedel-Crafts reaction with amounts of benzoyl chloride and aluminum chloride such that half of the polystyrene groups were converted to the benzophenone derivative. The product of this reaction was then treated with the Grignard reagent p-methoxyphenylmagnesium bromide to give the trityl alcohol derivative, which was in turn treated with acetyl chloride to give the trityl chloride derivative (Figure 3–6). The polystyrene methoxytrityl chloride product was isolated as a fluffy white powder. The final product chosen for oligonucleotide synthesis studies had 25 to 35 benzophenone groups per 100 styrene units, of which 5 to 7 had been converted to trityl chloride groups.

The first step in deoxyoligonucleotide synthesis was the attachment, through a covalent linkage, of a deoxynucleoside to the polymer derivative. The polymer was reacted with an excess of dry thymidine in pyridine, and methanol was added to convert any unreacted trityl chloride groups to the corresponding methyl ether. The large excess of nucleoside used made the protection of the 3' hydroxyl group unnecessary. On analysis it was found that about 15 per cent of the total trityl chloride groups had accepted thymidine, suggesting considerable steric hindrance around the trityl groups in the polymer. The thymidine linked to the polymer was stable to repeated precipitation of the polymer from aqueous medium and to alkaline treatment. Only by acetic acid treatment was thymidine released in the free form. These properties were taken as a good indication of covalent linkage to the polystyrene polymer. For inter-nucleotide bond synthesis 3'-O-acetylthymidine-5'-phosphate was reacted with the polymer-methoxytrityl-thymidine complex in the presence of mesitylenesulfonyl chloride, which served as the condensing agent. A large excess of the mononucleotide and condensing agent was used to maintain a concentration of activated phosphomonoester groups comparable to those used in previous step-wise syntheses. The condensation product, polystyrene-methoxytrityl-TpT, was isolated by aqueous precipitation. The yield was 93 per cent after two hours of reaction. Under similar reaction conditions N-benzoyl-3'-O-acetyladenosine-5'-phosphate, N-anisoyl-3'-O-acetylcytosine-5'-phosphate and N-benzoyl-3'-O-acetylguanosine-5'-phosphate were condensed with polystyrene-methoxytritylthymidine. The yields in all cases were nearly quantitative. The polymer with the attached dinucleotides was soluble in benzylamine, which was used to remove the protecting groups of the nucleotides. Methanolic ammonia or aqueous ammonia could not be used since these would have caused cleavage of the dinucleotide from the polymer.

Prior to oligonucleotide chain elongation, the acetyl group at the 3' hydroxyl end of the nucleotides was removed by incubation at room temperature for two minutes in a mixture of dimethylsulfoxide, pyridine and 1 M sodium methoxide in methanol. The nitrogen-protecting groups were completely stable for 10 minutes under these conditions.

Polystyrene-methoxytrityl-TpT was prepared by this method from the corresponding 3'-O-acetyl derivative, subjected to a repeat condensation with 3'-O-acetylthymidine-5'-phosphate and the product, polystyrene-methoxytrityl-TpTpT-O-acetyl, treated with alkali, precipitated, and then treated with acid to release the oligonucleotide material. The oligonucleotide product was analyzed by a combination of the techniques of paper chromatography and paper electrophoresis, and the yield at the second step of internucleotide bond synthesis was 88 per cent.

The advantages of the polymer support method are the high yields—between 88 and 96 per cent—and the easy purification of the products at each step by precipitation in aqueous solution. The method has potential for future syntheses but has not been further developed owing to interactions of the oligonucleotides with the support matrix and presumably because the defined oligonucleotides required for the coding studies and gene synthesis were synthesized by the other procedures.

Multiplication of Preformed Oligodeoxynucleotides and Synthesis of RNA of Complementary Sequence

The synthesis of RNA molecules of defined sequence was required for studies on genetic coding.[57] Step-wise synthetic methods had provided a series of oligodeoxynucleotides with repeating di- and trinucleotide sequences.[51,85,91] The enzyme DNA polymerase was used to prepare large amounts of polydeoxynucleotides for templates for RNA synthesis. DNA polymerase requires both complementary strands of the DNA primer for the enzyme catalyzed reaction to proceed. Thus, if the two oligodeoxynucleotides $d(TCC)_4$ and $d(AAG)_4$, which are base complementary with one another in the antiparallel Watson-Crick base pairing sense, are used as primers, the enzyme produces high molecular weight, faithfully replicated sequences of these short-chain templates.[12,54,56,134] The other pairs of repeating trinucleotides used are shown in Table 3–1.

To prepare single-stranded polyribonucleotides from the high molecular weight, double-stranded DNA polymers the enzyme DNA-dependent RNA polymerase was used. Because only two or three appropriate ribonucleoside triphosphate substrates were provided in the reaction mixture the action of the RNA polymerase was restricted to the transcription of only one strand of DNA at a time.[54,56,88] Using the example of poly[d(T-T-C)]•poly[d(G-A-A)],

$$\text{poly}[d(T\text{-}T\text{-}C)] \bullet \text{poly}[d(G\text{-}A\text{-}A)] + ATP + GTP \rightarrow \text{poly}[r(G\text{-}A\text{-}A)]$$

$$\text{poly}[d(T\text{-}T\text{-}C)] \bullet \text{poly}[d(G\text{-}A\text{-}A)] + UTP + CTP \rightarrow \text{poly}[r(U\text{-}U\text{-}C)]$$

Synthesis of RNA templates from the oligodeoxynucleotides listed in Table 3–1 yielded RNA's of defined sequence with sufficient triplet alternatives to verify the codon assignments for all 20 amino acids.[54,56]

Synthesis of Oligodeoxynucleotides and Joining Reactions

The various methods developed for synthesis of oligodeoxynucleotides have been used in the synthesis of the gene for yeast alanine transfer RNA.[2] This was accomplished by synthesis of oligodeoxynucleotides corresponding to sequences in the transfer RNA and to complementary sequences in the transfer RNA in such a manner that the sequences overlapped in the double-stranded configuration[55] (Figure 3–7). The synthesis of one of the oligonucleotides from Figure 3–7 is outlined in Figure 3–8. Other overlapping

ALANINE t-RNA
(NUCLEOTIDES 21–50)

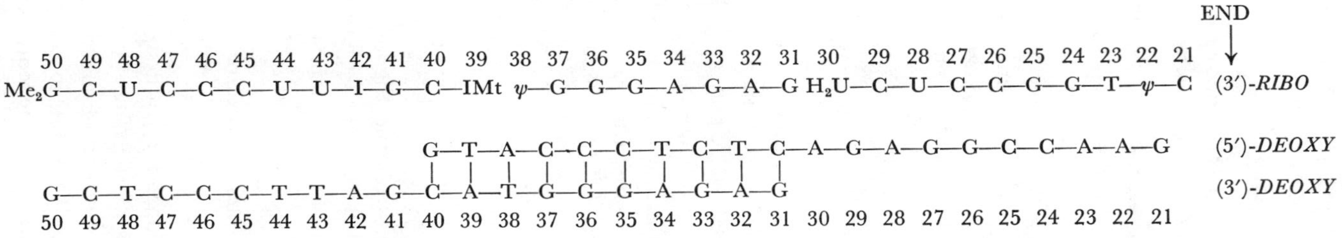

```
50  49 48 47 46 45 44 43 42 41 40 39 38 37 36 35 34 33 32 31 30 29 28 27 26 25 24 23 22 21      END
Me₂G—C—U—C—C—U—U—I—G—C—IMt ψ—G—G—G—A—G—A—G H₂U—C—U—C—C—G—T—ψ—C   (3')-RIBO

                                G—T—A—C—C—T—C—T—C—A—G—A—G—G—C—A—A—G    (5')-DEOXY
                                |   |   |   |   |   |   |   |   |
                                C—A—T—G—G—A—G—A—G
G—C—T—C—C—C—T—T—A—G—C—A—T—G—G—A—G—A—G                                   (3')-DEOXY
50  49 48 47 46 45 44 43 42 41 40 39 38 37 36 35 34 33 32 31 30 29 28 27 26 25 24 23 22 21
```

Figure 3-7. [From Khorana, H. G., Büchi, H., Caruthers, M. H., Chang, S. H., Gupta, N. K., Kumar, A., Ohtsuka, E., Sgaramella, V., and Weber, H., Cold Spring Harb. Symp. Quant. Biol. *33*, 35 (1968).]

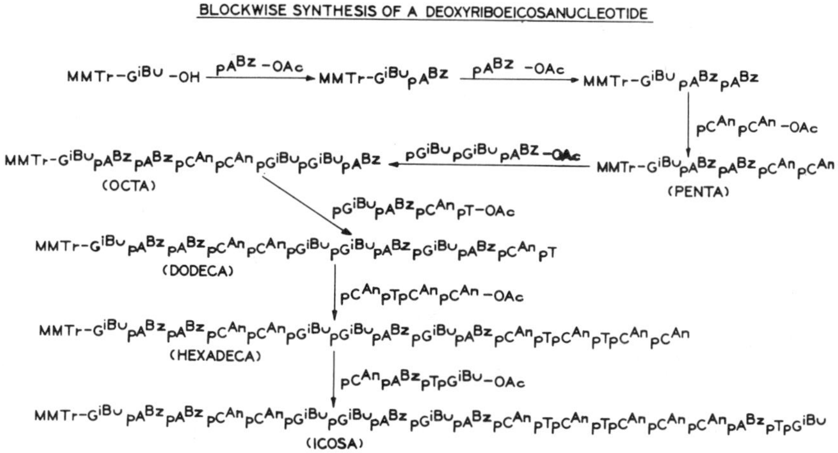

SYNTHESIS OF MODEL POLYNUCLEOTIDES / 87

BLOCKWISE SYNTHESIS OF A DEOXYRIBOEICOSANUCLEOTIDE

$MMTr-G^{iBu}-OH \xrightarrow{pA^{Bz}-OAc} MMTr-G^{iBu}pA^{Bz} \xrightarrow{pA^{Bz}-OAc} MMTr-G^{iBu}pA^{Bz}pA^{Bz}$

$\downarrow pC^{An}pC^{An}-OAc$

$MMTr-G^{iBu}pA^{Bz}pA^{Bz}pC^{An}pC^{An}pG^{iBu}pG^{iBu}pA^{Bz} \xleftarrow{pG^{iBu}pG^{iBu}pA^{Bz}-OAc} MMTr-G^{iBu}pA^{Bz}pA^{Bz}pC^{An}pC^{An}$
(OCTA) (PENTA)

$\downarrow pG^{iBu}pA^{Bz}pC^{An}pT-OAc$

$MMTr-G^{iBu}pA^{Bz}pA^{Bz}pC^{An}pC^{An}pG^{iBu}pG^{iBu}pA^{Bz}pG^{iBu}pA^{Bz}pC^{An}pT$
(DODECA)

$\downarrow pC^{An}pTpC^{An}pC^{An}-OAc$

$MMTr-G^{iBu}pA^{Bz}pA^{Bz}pC^{An}pC^{An}pG^{iBu}pG^{iBu}pA^{Bz}pG^{iBu}pA^{Bz}pC^{An}pTpC^{An}pTpC^{An}pC^{An}$
(HEXADECA)

$\downarrow pC^{An}pA^{Bz}pTpG^{iBu}-OAc$

$MMTr-G^{iBu}pA^{Bz}pA^{Bz}pC^{An}pC^{An}pG^{iBu}pG^{iBu}pA^{Bz}pG^{iBu}pA^{Bz}pC^{An}pTpC^{An}pTpC^{An}pC^{An}pA^{Bz}pTpG^{iBu}$
(ICOSA)

Figure 3-8. Steps in the chemical synthesis of an icosanucleotide. The standard method of presentation of polynucleotide chains is used. MMTr is the abbreviation for monomethoxytrityl group and is present at the 5'-OH of the terminal nucleoside. The protecting groups on the heterocyclic rings of different deoxynucleosides are shown by superscripts on the nucleoside initials; An stands for anisoyl, Bz for benzoyl and iBu for isobutyryl groups. OAc at the right-hand end of oligonucleotides stands for 3'-O-acetyl. [From Khorana, H. G., Büchi, H., Caruthers, M. H., Chang, S. H., Gupta, N. K., Kumar, A., Ohtsuka, E., Sgaramella, V., and Weber, H., Cold Spring Harb. Symp. Quant. Biol. **33**, 35 (1969).]

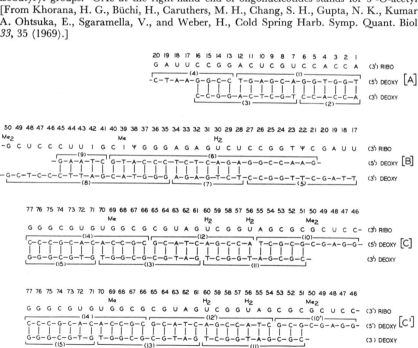

Figure 3-9. Total plan for the synthesis of a yeast alanine tRNA gene. The chemically synthesized segments are in brackets, the serial number of the segment being shown within the brackets. [From Agarwal, K. L., Büchi, H., Caruthers, M. H., Gupta, N., Khorana, H. G., Kleppe, K., Kumar, A., Ohtsuka, E., RajBhandary, U. L., van de Sande, J. H., Sgaramella, V., Weber, H., and Yamada, T., Nature **227**, 27 (1970).]

sequences were then synthesized, and eventually the entire gene was built up (Figure 3–9). The adjacent sequences were built up and joined sequentially. Joining was accomplished with the enzyme T4 polynucleotide ligase, which joins the 5' end of a 5'-phosphorylated oligonucleotide to the 3'-OH end of the adjacent oligonucleotide when the oligonucleotides are in bihelical complexes. A complete account of the gene synthesis is given in Agarwal, et al.[2]

STUDIES WITH SYNTHETIC MODEL POLYNUCLEOTIDES

The discovery that polynucleotide phosphorylase could be used to synthesize ribopolymers made available to nucleic acid chemists a series of ribonucleotide polymers of relatively well-defined sequence. These have provided the source material for a series of studies designed to investigate the structure and functioning of RNA from both the chemical and biological points of view. Nirenberg utilized synthetic ribopolymers as synthetic messenger RNA's in working out the triplet coding of RNA used in protein synthesis,[87] and numerous other workers have tested various other aspects of messenger function during protein synthesis. However, these biological functions are outside the scope of this monograph and will not be discussed. Studies on structure have been extended recently with the availability of deoxyribonucleotide synthetic polymers of absolutely defined sequence, synthesized by the techniques developed in Khorana's laboratory, or by use of the enzyme DNA terminal transferase.

Polymer-Polymer Interaction

The first structural studies using the synthetic polyribonucleotides developed from the observation, first described by Warner, of the interaction of ribopolymers with each other in solution. Warner showed that polyadenylic acid and polyuridylic acid interacted strongly on mixing.[131] This interaction was explored quantitatively by Davies, Felsenfeld and Rich, who mixed poly(A) and poly(U) in varying proportions and measured the hyperchromicity of the resulting mixtures.[20,98] They showed that in the absence of magnesium there was a 1 to 1 interaction of the two components, but in the presence of 1.2×10^{-2} M magnesium chloride the optimum interaction was shifted to a composition corresponding to 2 moles of uracil to 1 mole of adenine. On the basis of this observation these investigators suggested that in the absence of magnesium a poly(A)·poly(U) complex was formed and in the presence of magnesium a poly(A)·2 poly(U) complex was formed. Rich extended studies on the interaction of these ribopolymers by examining the X-ray diffraction patterns of the various interaction products of poly(A) and poly(U).[97] He showed that the interaction products had a helical configuration having the same general characteristics as DNA. This observation thus showed that the ribopolymers provided the necessary model compounds for investigation of secondary and tertiary structures of nucleic acids. The importance of these model compounds was so obvious that an intensive series of investigations on the interactions of poly(A) and poly(U), which was quickly extended to interactions between other synthetic ribopolymers, was undertaken. These studies have resulted in a large amount of information on the helix formation of ribopolymers of varying base sequences and the contribution of various types of base pairs to helix stabilization.[21,82,121]

In studies on helix stabilization the availability of synthetic polynucleotides of defined sequence has allowed investigation of the differential effects of the various bases, the effect of side-group substitutions on the sugar moieties, the effect on the stability of the helix of substitution of various side groups on the bases and, by differential studies, a comparison of the relative effects of various components in the helix stabilization phenomenon. These physical studies have in many instances provided additional circumstantial evidence to support various theories of biological function in terms of association of different types of molecules during biological processes.

Chamberlin and his collaborators have examined the thermal stability of a series of ribo- and deoxyribohomopolymer pairs.[14,15,101] They found that for a given base pair the thermal stability of the helix depends in a complicated manner on the sugar moieties of the two components of the helix—that is, whether the helix is composed of ribo- or deoxyribopolymers or mixed ribo-deoxyribopolymers. Thus, the thermal stability values vary in the orders dI:rC < dI:dC < rI:dC < rI:rC, dG:dC < dG:rC < rG:dC < rG:rC and rA:dT < dA:dT < rA:rT. In effect, all ribopolymer structures are more stable than all deoxyribopolymer structures or hybrid complexes of the same base pair. This conclusion is supported by more recent studies on poly(dU) complex formation.[137] To explain this phenomenon Chamberlin suggested that the configuration of the pentose ring may be involved.[101]. Ts'o, Rappaport and Bollum have suggested that hydrogen bonding of the 2' hydroxyl group to suitable sites, such as the 2 keto position of the pyrimidines or N-3 position of the purines, could explain most of the differences in physical properties and stabilities between ribo- and deoxyribopolymers and helix stability.[130] Brahms, Maurizot and Michelson have suggested that the ordered structures of RNA may be stabilized by formation of intermolecular hydrogen bonds to the nearby phosphate groups on the polymer backbone, a hypothesis based mainly on circular dichroism studies of dinucleotide model compounds.[9,10] Zmudzka, Janion and Shugar have examined the helix stabilization of the 2' hydroxyl group in poly(2'-O-methylcytidylic acid).[138] Poly(2'-O-methylcytidylic acid) forms a double-stranded helix under similar conditions to that formed by poly(rC), but not poly(dC), indicating that formation of the acid form of poly(rC) does not involve hydrogen bonding of the 2' hydroxyl to a base residue or to phosphate oxygen. They also showed that poly(2'-O-methyl C) readily complexes with poly(rI) to form a double stranded helix, the stability of which is less than that of poly(rI)·poly(rC) but greater than that of poly(rI)·poly(dC). Rottman studied the ribopolymer containing 2'-O-methyladenosine[103] and found that methylation of the 2' hydroxyl groups led to an increase in the thermal stability of the double-stranded polymer in acidic solution and also that methylation did not destabilize the double-stranded complex formed by poly(A) with poly(U).[6] On the basis of these studies and a comparison of dinucleotides containing 2'-O-methyl groups by nuclear magnetic resonance and circular dichroism spectroscopy Rottman suggested that hydrogen bonding involving 2' hydroxyl groups does not play a role in stabilization of ordered structures of polynucleotides, but rather that the important requirement for increased stability of polyribonucleotides compared to deoxyribonucleotides is the presence of an oxygen atom in the 2' position of the sugar.[7] The majority of studies on helix stability have involved an examination of the effect of different complementary pairs on helix stability and the effect of various side-group substitutions on these pairs. Inman and Baldwin studied the thermal stability of a number of deoxyhomopolymer pairs and showed that poly(dG)·poly(dC) has a much greater stability than poly(dI)·poly(dC) over a wide range of salt concentrations.[47,48] Szer and Shugar, studying ribohomopolymer interactions, showed that substitution of methyl groups in the pyrimidine 5 position increased the thermal stability of poly(rT)·poly(rA) and poly(r5-methyl-C)·poly(rI) complexes by about 20°C.[112,125] Studies from Michelson's laboratory and by Szer and Shugar on complexes of halogenated polymers with poly(rA) have shown that halogen substitution of bases increases thermal stability in the order $rU \simeq rFU < rClU < rBU \simeq rIU$.[75,124] The effect of the 5-halogen substitution compares favorably with the difference in helix stability of poly(rT)·poly(5-methyl-U) and poly (rU).[112] More recently Zmudzka, Bollum and Shugar have examined the thermal stability of poly(5-methyldeoxycytidylic acid) and some alkyl amino analogues.[136] In poly(5-methyldeoxycytidylic acid) the 5-methyl substituent does not have any appreciable effect on the thermal stability when complexed as an acid double-stranded helix. However, the poly(d5-methyl-C)·poly(dI) copolymer has a thermal stability 17°C higher than the poly(dI)·poly(dC) copolymer, confirming the significant stabilizing effect of a pyrimidine 5-methyl substituent even in the absence of a 2' hydroxyl group. The alkyl amino polymer analogues did not form double-stranded structures.

Interest in 5 substitution, particularly of 5'-methyl substitution, lies in the presence of 5-methyl substituted bases in naturally occurring nucleic acids. However, since the 5-methyl substitutions have not been indicated as the source of extra hydrogen bonds in the base-pair complexes, whether the stabilization effect is due to increased hydrogen bonding effects or to other factors is still open to question. Evidence on the contribution of hydrogen bonds to helix stability has been examined using poly(r2-aminoadenylic acid) complexes with poly(rU). Poly(r2-aminoadenylic acid) reacts with poly(rU) to form a 1 to 1 complex with three inter-base hydrogen bonds for each base pair.[42] The thermal stability is higher than the corresponding poly(rA)·poly(U) helix by 30°C. In similar mononucleotide polymer binding studies the thermal stability was increased about 15°C after introduction of 2-aminoadenylic acid and a third inter-base hydrogen bond.[43] Similar studies of a poly(dG)·poly(dC) homopolymer pair, which has three hydrogen bonds per base pair, to poly(dI)·poly(dC), with two hydrogen bonds per base pair, showed an increase in thermal stability of poly(dG)·poly(dC) over poly(dI)·poly(dC) complexes of about 50°C, suggesting either strong hydrogen bonds in the poly(dG)·poly(dC) pair or the presence of other base-specific stabilizing forces.[48]

Monomer and Oligonucleotide-Polymer Interaction

Most of the data quoted in the previous section have been obtained from polymer polymer interaction studies. However, a large amount of the research using synthetic polynucleotides as model compounds has involved monomer-polymer interactions or oligonucleotide-polymer interactions. The phenomenon of helix formation between monomers and polynucleotides was first reported by Howard, Frazier, Lipsett and Miles.[41] Studies in a number of laboratories have shown that the interactions between the monomers and polymers are in simple stoichiometric ratios and show specificity with respect to hydrogen bond base pairing just as at the polymer-polymer level. Also, the interactions are strongly cooperative, and the complexes formed have helix structures similar to those formed between corresponding pairs of homopolymers.[3,43,44,76] It is obvious that the use of monomeric components facilitates the study of base pairing. Since the interactions occur in aqueous solution, the results are more directly applicable to assessment of biological implications than observations of monomer-monomer interactions, which can only be carried out in organic solvents. Although a wide variety of purines have been shown to serve as monomeric components in these interactions, no pyrimidines form such complexes. Thus, only inferences by analogy can be made regarding pyrimidine components. Studies in Miles' laboratory on a variety of monomeric components with various side-group substituents have suggested that a necessary condition for complex formation is the presence of hydrogen bond donor and acceptor sites in locations appropriate for the formation of pairs of hydrogen bonds to the pyrimidine polymer.[43] The extent of protonation and pH effects revealed in these studies have aided understanding of the differences in complex formation of homopolymers at various pH's. There is a marked preference for three-stranded over two-stranded structures in monomer-polymer interactions, but substitution of a methyl group on the purine, or carrying out the reaction in a relatively high pH to limit protonation on half of the cytosine residues in the polymer, limits formation to two-stranded structures.[41] Huang and Ts'o, studying adenosine-poly(U) interactions, found that binding did not occur until a critical threshold concentration of adenosine was reached, a phenomenon which was salt-dependent.[44] On this basis they proposed that associated stacks of adenosine molecules serve as initiators for subsequent binding of individual adenosine molecules by a cooperative mechanism and that stacking of the molecules favors formation of monomer-polymer complexes. However, Felsenfeld and Miles have pointed out that the self-interaction of adenosine will be disrupted to form the poly(A)·2poly(U) helix.[21] Thus, its contribution will be negative since the orientation of the purines in the helical formation is fundamentally determined by hydrogen bonding; this factor, however, probably plays no significant role in self-association. These investigators point out that if the associated monomers have a similar orientation to those in the helices, they should show high optical rotation, but they do not do so. The initial studies in Miles' laboratory[43] and by Huang and Ts'o[44] utilizing infrared spectroscopy and optical rotation measurements, supported formation of a poly(A)·poly(U) complex at 20°C, compared to a poly(A)·2poly(U) complex at 5°C. However, Bangerter and Chan showed that binding of adenosine to polyuridylic acid occurs above 26°C by base stacking and intercalation only.[3] These studies utilized proton magnetic resonance spectroscopy. They also verified that a triple-stranded complex, stabilized by both adenine-uracil hydrogen bonding and adenine-adenine base stacking, is formed below this temperature.

In recent years there has been an interest in oligonucleotide-polymer interactions because of possible biological implications and biological mechanisms possibly involving such interactions. The original observations of oligonucleotide-polymer binding showed a dependence of the thermal stability on chain length.[71,72] Naylor and Gilham, examining the poly(dA)·poly(dT) series and the poly(rA)·poly(rU) series, showed that there was no interaction for chain lengths less than dT_5 and dA_4 and that the interaction was incomplete under their conditions for chain lengths less than dA_8 and dT_8.[86] Other workers have demonstrated reactions extending to lower chain lengths but under different conditions and higher concentrations of oligonucleotides. Cassani and Bollum have studied the interactions between a series of oligodeoxythymidylate oligonucleotides with polydeoxyadenylate and oligodeoxyadenylate oligonucleotides with polydeoxythymidylate.[13] They showed that the interactions of the various oligonucleotides with the polymers had very distinct thermal transition properties and that the reciprocal of the thermal stability temperature change was linear with the reciprocal of the chain length of the oligonucleotide. They also showed that oligodexoythymidylate oligonucleotides mixed in equal proportions with poly(dA) formed stable double-stranded complexes. With double the amount of oligonucleotide, triple-stranded complexes were formed which were less stable than the double-stranded ones. The interaction of deoxyadenylate oligonucleotides with poly(dT) was more complex. Triple-stranded structures involving proportions of two thymines with one adenine are formed, but none with proportions of two adenines with one thymine. These types of oligonucleotide polymer interactions have recently been extended to natural DNA's by Niyogi and Niyogi and Thomas, who have shown that there is a distinct difference in thermal stabilities in the binding of oligonucleotides of chain length 10 and longer to single-stranded naturally occurring DNA.[89,90] They have pointed out that this is the same length as one turn of the DNA helix.

Single-Stranded Polymer Interactions

A number of single-stranded synthetic polymers have the ability to form well-ordered secondary structures, and the configuration of single-stranded polynucleotides has been investigated by a number of laboratories.[21,82] Poly(rA) has been the most widely studied of the synthetic polymers in terms of homopolymer secondary structure, probably because at various pH's it shows different orientations. At pH 7 the polymer has properties which are affected by temperature,[25] but the hydrodynamic properties are those of a random coil[24,120]. In acidic solution of pH 5 the so-called acid structure of poly(A) possesses hydrodynamic properties that are characteristic of a rigid molecule,[24] and thermal effects indicate a loss of secondary structure over a narrow temperature range compared to the neutral poly(A) structure.[25,120,129] X-ray diffraction studies indicate a double-stranded structure with an inter-base distance of 3.8 Å.[23,79] A hydrogen-bonded double-stranded structure of poly(A)·poly(A) is possible with hydrogen bonds between the 6-amino group and the N-7 of the bases. The proton at N-1 is probably not involved in the bonding but may be involved in stabilizing the double helix by electrostatic interaction with phosphate groups.[23,99] Other evidence for the difference between the structures of poly(A) at neutral and acid pH is the reaction with formaldehyde. In the acid form the bases are protected from attack by formaldehyde, but the neutral form of poly(A) readily reacts with this reagent.[25] The original concept that the poly(rA) structure at neutral pH might consist of short, intra-molecularly bonded, looped regions has been shown to be incorrect by the studies with the ribopolymers 10-hydroxyethyladenylic acid and 10-dimethyladenylic acid, which do not allow hydrogen bonding of adenine and show the same temperature-dependent absorbance and optical rotatory properties as poly(rA).[28,40] Thermodynamic analysis of the structure has been investigated to obtain information of the stacking energy of the molecule. Van't Hoff analyses of denaturation data have shown that the standard free energy change obtained for single-stranded oligo- and polyadenylic acid at neutral pH and 0°C is about 1 kilocalorie per mole in favor of base stacking; this allows one in eight of the bases to be unstacked at this temperature, providing considerable flexibility to the molecule.[11,40,67,93,102] It has been calculated that at about 20°C two-thirds of the bases are in the stacked conformation at any given time.

Polycytidylic acid also forms an acid double-helical form.[35,128] However, the structure is somewhat different from that of poly(A). Thermal stability studies and the influence of pH and ionic strength, together with X-ray diffraction studies, have indicated that the hydrogen bonding system involves a shared proton between each pair of cytosine bases, the inter-base distance being 3.11 Å.[19,64] A similar structure is also formed by poly(dC).[1,46] At neutral pH poly(rC), like poly(rA), forms a disordered non-rigid structure.[19,35]

Poly(rU) forms no ordered structure at temperatures above ambient; however, a number of reports have appeared concerning the formation of secondary structure at low temperatures, but the nature of the ordered structure is still in doubt.[70,100,105]

Poly(G) aggregates very readily to form gels; because of this property evidence of the structure is difficult to obtain, and because of the difficulty in preparing poly(G) a number of conflicting reports have appeared in the literature with regard to its secondary structure. The only reliable preparations of poly(G) have been those using *E. coli* polynucleotide phosphorylase with manganese as co-factor.[127] Products of this reaction do possess a secondary structure, but the number of strands involved is still in doubt.[92]

Guanylic acid forms a gel in concentrated solution. X-ray diffraction studies of fibers drawn from such gels indicate stacking of tetramers with rotation about the helix axis.[45] The data obtained so far on poly(G) supports the concept that oligoguanylic acids aggregate readily and account for many of the difficulties in working with these compounds.

Polyinosinic acid forms a secondary structure consisting of three strands with an inter-base distance of 3.4 Å.[96] The thermal stability bears a linear relationship to the logarithm of the salt concentration. The structure formed by poly(dI) is slightly more stable than poly(I).[46] The optical rotatory dispersion spectrum of poly(I) is unusual in that the general spectrum of synthetic polyribonucleotides is reversed, with two troughs and one peak instead of two peaks and one trough. Formation of a complex with poly(C) or poly(A) inverts the optical rotatory dispersion profile.[104] On this basis it has been suggested that there may be a difference in the handedness of the helices. This concept is supported by recent X-ray diffraction studies on poly[d(I-C)]·poly[d(I-C)], a double-stranded molecule containing alternating sequences of deoxyriboinosinic and deoxyribocytidylic acids.[84] The molecule has an eight-fold helix (eight nucleotide pairs per turn) and is possibly left-handed.

Other recent studies on model polydeoxynucleotides of well-defined sequence show the influence of sequence on the properties of the molecules.[17,108,109,133] DNA's with the same base composition but different nucleotide sequence show differences in helix-coil transitions. DNA's with purines and pyrimidines on both strands are more stable than molecules with purines on one strand and pyrimidines on the complementary strand.[133]

As more polymers of known sequence become available the finer aspects of sequence effects on polynucleotide structure will be understood, allowing better understanding of the structure-function relationships of DNA, RNA and protein.

REFERENCES

1. Adler, A., Grossman, L., and Fasman, G. D., Proc. Nat. Acad. Sci. (Wash.) 57, 423 (1967).
2. Agarwal, K. L., Büchi, H., Caruthers, M. H., Gupta, N., Khorana, H. G., Kleppe, K., Kumar, A., Ohtsuka, E., RajBhandary, U. L., van de Sande, J. H., Sgaramella, V., Weber, H., and Yamada, T., Nature 227, 27 (1970).
3. Bangerter, B. W., and Chan, S. I., Proc. Nat. Acad. Sci. (Wash.) 60, 1144 (1968).
4. Beers, R. F., Jr., Arch. Biochem. Biophys. 75, 497 (1958).
5. Bernfield, M. R., J. Biol. Chem. 241, 2014 (1966).
6. Bobst, A. M., Cerutti, P. A., and Rottman, F., J. Amer. Chem. Soc. 91, 1246 (1969).
7. Bobst, A. M., Rottman, F., and Cerutti, P. A., J. Amer. Chem. Soc. 91, 4603 (1969).
8. Bollum, F. J., Groeniger, E., and Yoneda, H., Proc. Nat. Acad. Sci. (Wash.) 51, 853 (1964).
9. Brahms, J., Maurizot, J. C., and Michelson, A. M., J. Mol. Biol. 25, 465 (1967).
10. Brahms, J., Maurizot, J. C., and Michelson, A. M., J. Mol. Biol. 25, 481 (1967).
11. Brahms, J., Michelson, A. M., and van Holde, K. E., J. Mol. Biol. 15, 467 (1966).
12. Byrd, C. E., Ohtsuka, E., Moon, M. W., and Khorana, H. G., Proc. Nat. Acad. Sci. (Wash.) 53, 79 (1965).
13. Cassani, G. R., and Bollum, F. J., Biochemistry 8, 3928 (1969).
14. Chamberlin, M. J., Fed. Proc. 24, 1446 (1965).
15. Chamberlin, M. J., and Patterson, D., J. Mol. Biol. 12, 410 (1965).
16. Cramer, F., Helbig, R., Hettler, H., Scheit, K. H., and Seliger, H., Angew. Chem. 78, 640 (1966).
17. Elson, E. L., Scheffler, I. E., and Baldwin, R. L., J. Mol. Biol. 54, 401 (1970).
18. Entner, N., and Gonzalez, C., Biochem. Biophys. Res. Commun. 1, 333 (1959).
19. Fasman, G D , Lindblow, C., and Grossman, L., Biochemistry 3, 1015 (1964).
20. Felsenfeld, G., Davis, D. R., and Rich, A., J. Amer. Chem. Soc. 79, 2023 (1957).
21. Felsenfeld, G., and Miles, H. T., Ann. Rev. Biochem. 36, 407 (1967).
22. Fitt, P. S., and See, Y. P., Biochem. J., 116, 309 (1970).
23. Fresco, J. R., J. Mol. Biol. 1, 106 (1959).
24. Fresco, J. R., and Doty, P., J. Amer. Chem. Soc. 79, 3928 (1957).
25. Fresco, J. R., and Klemperer, E., Ann. N. Y. Acad. Sci. 81, 730 (1959).
26. Gilham, P. T., and Khorana, H. G., J. Amer. Chem. Soc. 80, 6212 (1958).
27. Gilham, P. T., and Khorana, H. G., J. Amer. Chem. Soc. 81, 4647 (1959).
28. Griffin, B. E., Haslam, W. J., and Reese, C. B., J. Mol. Biol. 10, 353 (1964).
29. Griffin, B. E., Todd, A., and Rich, A., Proc. Nat. Acad. Sci. (Wash.) 44, 1123 (1958).
30. Grunberg-Manago, M., in *Acides Ribonucléiques et Polyphosphates*, Colloq. Intern. du CNRS, Strasbourg, 1961. CNRS, Paris, p. 295 (1962).
31. Grunberg-Manago, M., Prog. Nucleic Acid Res. Mol. Biol. 1, 93 (1963).
32. Grunberg-Manago, M., and Ochoa, S., J. Amer. Chem. Soc. 77, 3165 (1955).
33. Grunberg-Manago, M., Ortiz, P. J., and Ochoa, S., Biochim. Biophys. Acta 20, 269 (1956).
34. Hayatsu, H., and Khorana, H. G , J. Amer. Chem. Soc. 89, 3880 (1967).
35. Helmkamp, G. K., and Ts'o, P. O. P., Biochim, Biophys. Acta 55, 601 (1962).
36. Hendley, D. D., and Beers, R. F., Jr., J. Biol. Chem. 236, 2050 (1961).
37. Heppel, L. A., Singer, M. F., and Hilmoe, R. J., Ann. N.Y. Acad. Sci. 81, 635 (1959).

38. Heppel, L A., Whitfield, P. R., and Markham, R., Biochem. J. *60*, 8 (1955).
39. Hilmoe, R. J., and Heppel, L. A., J. Amer. Chem. Soc. *79*, 4810 (1957).
40. van Holde, K. E., Brahms, J., and Michelson, A. M., J. Mol. Biol. *12*, 726 (1965).
41. Howard, F. B., Frazier, J., Lipsett, M. N., and Miles, H. T., Biochem. Biophys. Res. Commun. *17*, 93 (1964).
42. Howard, F. B., Frazier, J., and Miles, H. T., J. Biol. Chem. *241*, 4293 (1966).
43. Howard, F. B., Frazier, J., Singer, M. F., and Miles, H. T., J. Mol. Biol. *16*, 415 (1966).
44. Huang, W. M., and Ts'o, P. O. P., J. Mol. Biol. *16*, 523 (1966).
45. Iball, J., Morgan, C. H., and Wilson, H. R., Nature *199*, 688 (1963).
46. Inman, R. B., J. Mol. Biol. *9*, 624 (1964).
47. Inman, R. B., and Baldwin, R. L., J. Mol. Biol. *5*, 172 (1962).
48. Inman, R. B., and Baldwin, R. L., J. Mol. Biol. *8*, 452 (1964).
49. Jacob, T. M., and Khorana, H. G., J. Amer. Chem. Soc. *86*, 1630 (1964).
50. Jacob, T. M., and Khorana, H. G., J. Amer. Chem. Soc. *87*, 368 (1965).
51. Jacob, T. M., and Khorana, H. G., J. Amer. Chem. Soc. *87*, 2971 (1965).
52. Janion, C., Zmudzka, B., and Shugar, D., Acta Biochim. Polon. *17*, 31 (1970).
53. Keir, H. M., and Smith, M. J., Biochim. Biophys. Acta *68*, 589 (1963).
54. Khorana, H. G., in *Genetic Elements: Properties and Function* (D. Shugar, Ed.), Academic Press, New York, p. 209 (1967).
55. Khorana, H. G., Büchi, H., Caruthers, M. H., Chang, S. H., Gupta, N. K., Kumar, A., Ohtsuka, E., Sgaramella, V., and Weber, H., Cold Spring Harb. Symp. Quant. Biol. *33*, 35 (1968).
56. Khorana, H. G., Büchi, H., Ghosh, H., Gupta, N., Jacob, T. M., Kössel, H., Morgan, R., Narang, S. A., Ohtsuka, E., and Wells, R. D., Cold Spring Harb. Symp. Quant. Biol. *31*, 39 (1966).
57. Khorana, H. G., Büchi, H., Jacob, T. M., Kössel, H., Narang, S. A., and Ohtsuka, E., J. Amer. Chem. Soc. *89*, 2154 (1967).
58. Khorana, H. G., Razzell, W. E., Gilham, P. T., Tener, G. M., and Pol. E. H., J. Amer. Chem. Soc. *79*, 1002 (1957).
59. Khorana, H. G., Tener, G. M., Moffat, J. G., and Pol. E. H., Chemistry and Industry (London) 1523 (1956).
60. Khorana, H. G., and Vizsolyi, J. P., J. Amer. Chem. Soc. *83*, 675 (1961).
61. Klee, C. B., and Singer, M. F., Biochem. Biophys. Res. Commun. *29*, 356 (1967).
62. Klee, C. B., and Singer, M. F., J. Biol. Chem. *243*, 923 (1968).
63. Krakow, J. S., Coutsogeorgopoulos, C., and Canellakis, E. S., Biochem. Biophys. Res. Commun. *5*, 477 (1961).
64. Langridge, R., and Rich, A., Nature *198*, 725 (1963).
65. Lee-Huang, S., and Cavalieri, L. F., Proc. Nat. Acad. Sci. (Wash.) *50*, 1116 (1963).
66. Lee-Huang, S., and Cavalieri, L. F., Proc. Nat. Acad. Sci. (Wash.) *51*, 1022 (1964).
67. Leng, M., and Felsenfeld, G., J. Mol. Biol. *15*, 455 (1966).
68. Lengyel, P., and Chambers, R. W., J. Amer. Chem. Soc. *82*, 752 (1960).
69. Levin, D. H., Biochim, Biophys. Acta *61*, 75 (1962).
70. Lipsett, M. N., Proc. Nat. Acad. Sci. (Wash.) *46*, 445 (1960).
71. Lipsett, M. N., J. Biol. Chem. *239*, 1256 (1964).
72. Lipsett, M. N., Heppel, L. A., and Bradley, D. F., J. Biol. Chem. *236*, 857 (1961).
73. Littauer, U. Z., and Kornberg, A., J. Biol. Chem. *226*, 1077 (1957).
74. Lohrmann, R., Söll, D., Hayatsu, H., Ohtsuka, E., and Khorana, H. G., J. Amer. Chem. Soc. *88*, 819 (1966).
75. Massoulié, J., Michelson, A. M., and Pochon, F, Biochim. Biophys. Acta *114*, 16 (1966).
76. Maxwell, E. S., Barnett, L., Howard, F. B., and Miles, H. T., J. Mol. Biol. *16*, 440 (1966).
77. Merrifield, R. B., Science *150*, 178 (1965).
78. Michelson, A. M., Nature *181*, 303 (1958).
79. Michelson, A. M., J. Chem. Soc. 1371 (1959).
80. Michelson, A. M., Dondon, J., and Grunberg-Manago, M., Biochim. Biophys. Acta *55*, 529 (1962).
81. Michelson, A. M., and Grunberg-Manago, M., Biochim. Biophys. Acta *91*, 92 (1964).
82. Michelson, A. M., Massoulié, J., and Guschlbauer, W., Prog. Nucleic Acid Res. Mol. Biol. *6*, 84 (1967).
83. Michelson, A. M., and Todd, A., J. Chem. Soc. 2632 (1955).

REFERENCES / 97

84. Mitsui, Y., Langridge, R., Shortle, B. E., Cantor, C. R., Grant, R. C., Kodama, M., and Wells, R. D., Nature *228*, 1166 (1970).
85. Narang, S. A., Jacob, T. M., and Khorana, H. G., J. Amer. Chem. Soc. *89*, 2158 (1967).
86. Naylor, R., and Gilham, P. T., Biochemistry *5*, 2722 (1966).
87. Nirenberg, M., Caskey, T., Marshall, R., Brimacombe, R., Kellogg, D., Doctor, B., Hatfield, D., Levin, J., Rottman, F., Pestka, S., Wilcox, M., and Anderson, F., Cold Spring Harb. Symp. Quant. Biol. *31*, 11 (1966).
88. Nishimura, S., Jones, D. S., and Khorana, H. G., J. Mol. Biol. *13*, 302 (1965).
89. Niyogi, S. K., J. Biol. Chem. *244*, 1576 (1969).
90. Niyogi, S. K., and Thomas, C. A., Jr., Biochem. Biophys. Res. Commun. *26*, 51 (1967).
91. Ohtsuka, E., Moon, M. W., and Khorana, H. G., J. Amer. Chem. Soc. *87*, 2956 (1965).
92. Pochon, F., and Michelson, A. M., Proc. Nat. Acad. Sci. (Wash.) *53*, 1425 (1965).
93. Poland, D., Vournakis, J. N., and Scheraga, H. A., Biopolymers *4*, 223 (1966).
94. Radding, C. M., and Kornberg, A., J. Biol. Chem. *327*, 2877 (1962).
95. Ralph, R. K., and Khorana, H. G., J. Amer. Chem. Soc. *83*, 2926 (1961).
96. Rich, A., Biochim. Biophys. Acta *29*, 502 (1958).
97. Rich, A., Brookhaven Symp. Biol. *12*, 17 (1959).
98. Rich, A., and Davies, D. R., J. Amer. Chem. Soc. *78*, 3548 (1956).
99. Rich, A., Davies, D. R., Crick, F. H. C., and Watson, J. D., J. Mol. Biol. *10*, 28 (1961).
100. Richards, E. G., Flessel, C. P., and Fresco, J. R., Biopolymers *1*, 431 (1963).
101. Riley, M., Maling, B., and Chamberlin, M. J., J. Mol. Biol. *20*, 359 (1966).
102. Ross, P. D., and Scruggs, R. L., Biopolymers *3*, 491 (1965).
103. Rottman, F., and Heinlein, K., Biochemistry *7*, 2634 (1968).
104. Sarkar, P. K., and Yang, J. T., Biochem *4*, 1238 (1965).
105. Sarkar, P. K., and Yang, J. T., J. Biol. Chem. *240*, 2088 (1965).
106. Schachman, H. K., Adler, J., Radding, C. M., Lehmann, I., and Kornberg, A., J., Biol. Chem. *235*, 3242 (1960).
107. Schaller, H., and Khorana, H. G., J. Amer. Chem. Soc. *85*, 3828 (1963).
108. Scheffler, I. E., Elson, E. L., and Baldwin, R. L., J. Mol. Biol. *36*, 291 (1968).
109. Scheffler, I. E., and Sturtevant, J. M., J. Mol. Biol. *42*, 477 (1969).
110. See, Y. P., and Fitt, P. S., Biochem. J. *119*, 517 (1970).
111. Shemyakin, M. M., Ovchinnikov, Y. A., Kinyushkin, A. A., and Kozhevnikova, I. V., Tetrahedron Letters 2323 (1965).
112. Shugar, D., and Szer, W., J. Mol. Biol. *5*, 580 (1962).
113. Singer, M. F., J. Biol. Chem. *238*, 336 (1963).
114. Singer, M. F., and Guss, J. K., J. Biol. Chem. *237*, 182 (1962).
115. Singer, M. F., Heppel, L. A., and Hilmoe, R. J., Biochem. Biophys. Acta *26*, 447 (1957).
116. Singer, M. F., Heppel, L. A., and Hilmoe, R. J., J. Biol. Chem. *235*, 738 (1960).
117. Singer, M. F., Heppel, L. A., Hilmoe, R. J., Ochoa, S., and Mii, S., Can. Cancer Conf. *3*, 41 (1959).
118. Singer, M. F., and O'Brien, B. M., J. Biol. Chem. *238*, 328 (1963).
119. Smith, M. J., and Khorana, H. G., J. Amer. Chem. Soc. *81*, 2911 (1959).
120. Steiner, R. F., and Beers, R. F., Biochim. Biophys. Acta *26*, 336 (1957).
121. Steiner, R. F., and Beers, R. F., *Polynucleotides*, Elsevier Publishing Company, Amsterdam, p. 235 (1961).
122. Swierkowski, M., and Shugar, D., J. Mol. Biol. *47*, 57 (1970).
123. Szer, W., and Shugar, D., Acta Biochim. Polon. *8*, 235 (1961).
124. Szer, W., and Shugar, D., Acta Biochim. Polon. *10*, 219 (1963).
125. Szer, W., and Shugar, D., J. Mol. Biol. *17*, 174 (1966).
126. Tener, G. M., Khorana, H. G., Markham, R., and Pol, E. H., J. Amer. Chem. Soc. *80*, 6224 (1958).
127. Thang, M. N., Graffe, M., and Grunberg-Manago, M., Biochim. Biophys. Acta *108*, 125 (1965).
128. Ts'o, P. O. P., Helmkamp, G. K., and Sander, C., Biochim. Biophys. Acta *55*, 584 (1962).
129. Ts'o, P. O. P., Helmkamp, G. K., and Sander, C., Proc. Nat. Acad. Sci. (Wash.) *48*, 686 (1962).
130. Ts'o, P. O. P., Rapaport, S. A., and Bollum, F. J., Biochemistry *5*, 4153 (1966).

131. Warner, R. C., J. Biol. Chem. *229*, 711 (1957).
132. Weimann, G., Schaller, H., and Khorana, H. G., J. Amer. Chem. Soc. *85*, 3835 (1963).
133. Wells, R. D., Larson, J. E., Grant, R. C., Shortle, B. E., and Cantor, C. R., J. Mol. Biol. *54*, 465 (1970).
134. Wells, R. D., Ohtsuka, E., and Khorana, H. G., J. Mol. Biol. *14*, 221 (1965).
135. Yoneda, M., and Bollum, F. J., J. Biol. Chem. *240*, 3385 (1965).
136. Zmudzka, B. Bollum, F. J., and Shugar, D., Biochemistry *8*, 3049 (1969).
137. Zmudzka, B., Bollum, F. J., and Shugar, D., J. Mol. Biol. *46*, 169 (1969).
138. Zmudzka, B., Janion, C., and Shugar, D., Biochem. Biophys. Res. Commun. *37*, 895 (1969).
139. Zmudzka, B., and Shugar, D., FEBS Letters *8*, 52 (1970).

CHAPTER 4   PHYSICAL METHODOLOGY AND THE MACROMOLECULAR PROPERTIES OF THE NUCLEIC ACIDS

Part of the challenge of research work on nucleic acids is the fact that they are high molecular weight polyelectrolytes. This property has made the study of their macromolecular properties particularly difficult and has resulted in the development and modification of many specialized techniques. Physical methods used to study macromolecular properties of the nucleic acids are based mainly on the hydrodynamic and spectroscopic properties of the molecules. Most of the information obtained pertains to structure—for example, size, shape, mass and inter- and intra-molecular interactions.

Discussions of certain physical methods, such as X-ray diffraction and light scattering techniques, have purposefully not been included in this chapter. X-ray diffraction is not an easy technique, nor is it generally available, being limited to a few laboratories having the necessary equipment and a staff with the technical expertise. Almost every textbook describes the technique, and reviews appear regularly.[25,63,125] Light scattering, an important technique in protein chemistry, is not applicable to nucleic acids having molecular weights above 3,000,000 daltons;[32] it also is not a generally available technique.

Osmotic Pressure

Osmotic pressure is a measure of the movement of solvent (and low molecular weight solutes if present) across a semi-permeable membrane, one which is impermeable to high molecular weight solutes.

A schematic diagram of an osmometer is shown in Figure 4–1. Passage of solvent from side I to side II causes the liquid level of side II to rise in the capillary tube. This flow continues until the concentrations on both sides of the membrane are equal. However, the rise of liquid in the capillary tube sets up a hydrostatic pressure which opposes the flow of solvent through the membrane, and an equilibrium is attained. The pressure difference between the two sides at equilibrium is the osmotic pressure.

Osmotic pressure is a colligative property of solutions; that is, it provides a measure of the chemical potential of the solvent in a solution relative to that of pure solvent, and this permits the determination of the molecular weight of the solute. Osmotic pressure for the nucleic acids is an example of the type of limitation imposed by the size of a molecule. It cannot be used in the study of the high molecular weight nucleic acids since changes in osmotic pressure become so minute when the particles being studied have molecular weights above 1,000,000 daltons; however, it is used widely in the study of proteins.

The relationship between osmotic pressure and molecular weight is given by van't Hoff's limiting law for osmotic pressure:

$$\lim_{C \to 0} \frac{\Pi}{C} = \frac{RT}{M} \tag{1}$$

where C, the concentration of solute, is extrapolated to zero concentration to approximate the ideal state, M is the molecular weight of the solute, Π the osmotic pressure, R the gas constant and T the temperature. At zero concentration $\bar{V} = V°$, where $\bar{V}$ is the partial molal volume of the solvent and $V°$ is the molal volume of the pure solvent. Since any practical measurement deals with a population of macromolecules, unless the preparations are homogeneous, the number of macromolecules must be considered and the equation then gives $\bar{M}$, the number average molecular weight:

$$\lim_{C \to 0} \frac{\Pi}{C} = \frac{RT}{\bar{M}} \tag{2}$$

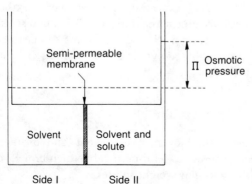

Figure 4–1. Schematic diagram of an osmometer.

These equations have been defined for macromolecular non-electrolytic solutes. The equations hold true for polyelectrolytes as well as for neutral macromolecules, provided that certain experimental parameters, such as ionic strength, are maintained constant during any set of measurements. This can be accomplished by dilution procedures in which the total concentration of counterions is kept constant, so-called isoionic dilution.

One parameter not considered in the equations is the Donnan effect.[29] In a solution of polyelectrolyte any small ions permeable to the membrane will be unequally distributed on the two sides, resulting in a pH difference. However, the relative concentration differences on either side of the semipermeable membrane are decreased if the total electrolyte concentration is sufficiently high. Thus, the Donnan effect can be supressed in the presence of a sufficiently high concentration of neutral salt, and pH difference will be effectively eliminated by the presence of, for example, an excess of sodium chloride in a polyelectrolyte solution.

As mentioned previously, osmotic pressure has been used to determine the molecular weight of many proteins. The technique may have some application to relatively small nucleic acids provided that care is exercised with respect to ionic concentration and conformation and that the results are compared with other methods of analysis, criteria basic to all physical methods applied to nucleic acids.

HYDRODYNAMIC PROPERTIES

Viscosity

An extremely important hydrodynamic property of macromolecules is viscosity. Viscosity of nucleic acids has been studied extensively and used in a large number of molecular weight determinations. The large size and conformation of nucleic acid molecules in solution and their polyelectrolyte nature has necessitated modifications and particular adaptations of the classic approach to viscosity and viscosity measurements.

The original definition of viscosity by Newton considers a viscous fluid one in which there are attractive forces between neighboring portions of the fluid. These attractive forces oppose any motion of one part of the fluid relative to another part. From a theoretical point of view the liquid was considered to be a structureless, continuous fluid. When two adjacent volume elements are considered, in which one volume element is in motion relative to the other, the forces opposing the motion are called the frictional force. The frictional force is proportional to the relative velocity and to the contact area between the adjacent volume elements; it is inversely proportional to the distance between the centers of the elements. The proportionality constant relating this force to the variables is the coefficient of viscosity. Thus,

$$F = \eta \left(\frac{du}{dx}\right) dA \tag{3}$$

where F is the frictional force, η the coefficient of viscosity, du the relative velocity, dx the distance between the centers of the elements and dA the area between the adjacent volume elements.

Considering liquids in motion, the frictional (viscous) force prevents very abrupt changes in velocity with distance, since the larger the expression du/dx, the greater the force tending to decrease it. This property raises two important considerations. If we consider a liquid flowing along a tube, the wall of the tube is, by definition, stationary and attractive forces exist between the material in the wall of the tube and the liquid. Thus, the liquid immediately adjacent to the wall will have a very low velocity. Liquid slightly farther from the wall will have a slightly higher velocity, since the attractive forces between the elements of the liquid are less. This results in velocity gradients throughout the tube with the maximum velocity at the center of the tube. Thus, in calculating total resistance to flow the frictional forces existing at all points in the tube must be taken into account. Since the force at the wall is a negligible portion of the whole, the viscosity measurement of η is taken as the viscosity of the liquid. These concepts apply to the measurement of viscosity by the capillary and rotating cylinder viscometers which will be described later.

The second consideration concerns a solid particle moving through a stationary liquid. The elements of fluid adjacent to the particle will move with approximately the same velocity as the particle. Those farther away, with a slightly lower velocity, and liquid will be in motion up to a considerable distance away from the particle. Thus, the total frictional force will be largely the friction between adjacent elements of the liquid. Ignoring the small forces adjacent to the particle, we may consider the resistance to motion as independent of the material of which the particle is made, depending only on the viscosity of the liquid. This principle is important in measurement of viscosity by the falling ball method. Calculation of the resisting force, or frictional coefficient, involves a computation of the pattern of liquid flow in the region around the particle followed by integration of equation 3 over the volume of all elements in that region. The problem was solved for spherical particles by Stokes[109] and for ellipsoids of revolution by Perrin,[88] also Herzog, Illig and Kudar.[47] To make this calculation it is necessary to derive a three-dimensional equation of movement for fluid flow of the type mass × acceleration = force, per unit volume of fluid. Solution of this equation can be used to calculate the total force on the immersed particle, a force which consists of two parts—the force due to the variation of the hydrostatic pressure with position and the frictional force. Both of these parts are proportional to the coefficient of viscosity and to the radius, r, of the particle. Thus,

$$F = 6\pi\eta r u \qquad (4)$$

where F is the frictional force, r the radius of the particle and u the velocity, which is constant. This is identical to Stokes' law for the frictional coefficient of a sphere moving in a fluid of viscosity η:

$$f = 6\pi\eta r \qquad (5)$$

where f is the frictional coefficient and r the radius of the sphere. It was from Stokes' equation that Perrin and Herzog, Illig and Kudar extended the equations to ellipsoids of revolution, both for prolate ellipsoids and oblate ellipsoids. Stokes' equation has subsequently been confirmed by measurement of the sedimentation coefficients of spherical polystyrene particles whose radii were measured by electron microscopy.[16]

The equations are all based on random orientation of the molecules in the fluid, which occurs when the flow velocity is sufficiently small. For macromolecular solutions this is particularly important since the frictional force and frictional coefficient of asymmetric particles depends on their orientation with respect to the direction of flow. Thus, any ordering of the molecules in the fluid must be avoided. For some solutions, particularly macromolecular solutions, the phenomenon of flow itself causes orientation of the constituent molecules and at high velocity deformation of the molecules; such liquids are said to be non-Newtonian. If non-Newtonian behavior cannot be avoided, Newtonian viscosity—viscosity independent of velocity gradient—can be calculated when the shearing forces are small by extrapolation to zero velocity gradient.

Fluid flowing slowly along a tube without obstacles has a pattern of flow which, when observed experimentally, is streamline or laminar flow. At high velocities or near obstacles this pattern of flow may be disturbed, vortices may appear and the fluid will move both with velocity in the direction of motion and with rotational velocity. Such disturbed flow is called turbulent flow. In order to measure viscosity accurately the flow must be laminar, since the flow pattern for turbulent flow is extremely complex. Laminar flow is easily obtained at low velocities in narrow capillaries or between closely opposed slowly rotating cylinders. These two types of apparatus are the basic features of the usual type of viscometers.

Capillary Viscometers

When a fluid flows through a uniform capillary of radius r and length l under the influence of a pressure of P dynes per square centimeter, the thin layer of liquid in contact with the walls of the capillary will be approximately stationary, as already described. The flow of the adjacent layer of liquid will be slowed by viscosity, but the effect of viscosity will continue to a diminishing extent to the center of the tube. Thus, the rate of flow of the liquid will become less the smaller the radius of the tube. The correlation of these factors was first derived by Poiseuille, who showed that a liquid having a coefficient of viscosity η flowing at uniform velocity at a rate of v cubic centimeters per second could be described by the equation

$$\eta = \frac{\pi P r^4}{8vl} \tag{6}$$

Utilization of the principles of this equation is the basic principle on which the capillary viscometer is based.[13,121] In the Ostwald-type capillary viscometer (Figure 4–2) a known volume of liquid is measured in terms of time of its flow between two marks (l) on the capillary tube. At any instant the pressure driving the liquid through the capillary is equal to $hg\rho$, where h is the difference in height between the levels of the liquid in the two limbs of the U-shaped capillary tube. This value varies during the experiment, but as the initial and final values are the same in every instance, the applied pressure is thus proportional to the density, ρ, of the liquid. Since the same capillary tube is used for more than one experiment, and measurements are done with both the solvent and the solute, r and l are constant and the same volume of liquid flows through the capillary. In Poiseuille's original equation the rate of flow is a function of volume and time, so that the expression can be written

$$\eta = \frac{\pi P r^4 t}{8vl} \tag{7}$$

where the rate of flow is v cubic centimeters per t seconds. Thus, for two liquids 1 and 2

$$\frac{\eta_1}{\eta_2} = \frac{\rho_1 t_1}{\rho_2 t_2} \tag{8}$$

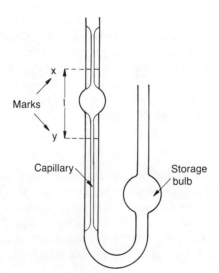

Figure 4–2. Ostwald-type viscometer.

where t_1 and t_2 are the times of flow. Since one liquid is solvent of known viscosity, equation 8 gives the viscosity of the solute. Poiseuille's law ignores the fact that the liquid issuing from the capillary has greater kinetic energy than that which entered the capillary. For slow flow velocities this difference is negligible and can be ignored, but for viscometers with short flow times—that is, larger velocities—a correction must be applied.

Rotating Cylinder Viscometers

A major problem encountered in the measurement of macromolecular solutions, such as nucleic acids, with capillary-type viscometers is the high shear stress of the apparatus. Specially designed capillary viscometers have been constructed,[46] but the shear stress is too high for study of high molecular weight nucleic acids, particularly DNA.[11,133] With very large and stiff molecules, such as DNA, the viscosity decreases as the shear stress increases, and the thermal motions of the molecules in solution are slow compared to the orientations induced by flow in a capillary. This gives rise to a non-Newtonian situation, which, as described previously, must be avoided. Low shear viscometers of the rotating cylinder-type eliminate this constraint upon the measurement of the viscosity of macromolecules[34] and were first described by Couette.[21] Rotating cylinder viscometers consist of two concentric cylinders, one rotating and one at rest. The flow lines are circular with the velocity of the liquid zero at the surface of the resting cylinder and increasing to a maximum at the surface of the rotating cylinder. The viscosity of the liquid is linearly related to the torque exerted on the resting cylinder at a given speed of rotation, and the speed of rotation of the rotor is inversely proportional to the viscosity of the liquid to be measured. Since the rotor speed, and thus the shear stress, can be varied, measurements can be made at various shear stresses. As in the capillary viscometer readings are taken on the liquid to be measured and on the solvent, and the relative viscosity is determined by the time for the rotor to rotate a given number of revolutions divided by the time for the rotor to rotate the same number of revolutions in pure solvent. A rotating cylinder viscometer designed by Zimm and Crothers[133] allows the apparatus to be maintained at constant temperature and bacteriophage DNA has been measured at a shear stress as low as 0.0006 dyne per square centimeter.[24] This compares with a shear stress of about 10 dynes per square centimeter, obtained with a capillary viscometer of flow volume of about 1 milliliter and a time of flow of about 100 seconds for water. Zimm[132] has calculated that a capillary viscometer which would provide a shear stress of about 0.1 dyne per square centimeter would have to be 10 meters long and have a flow time of 10,000 seconds (approximately three hours) for a 1 milliliter volume. Even at 0.1 dyne per square centimeter the shear stress would still be too high for accurate measurements of DNA molecules. The commercially available rotating cylinder viscometer based on the design of Zimm and Crothers[133]

Falling Sphere Viscometers

The third method of measuring viscosity is the falling sphere method. This has rarely been used with nucleic acid solutions because of the large volume of solution required. The apparatus consists of a cylinder of solution and a sphere of known density and dimensions. The viscosity is determined from the time required for the sphere to fall between two marks on the cylinder in both solute and solvent. If a sphere is falling under the influence of gravity, the constant downward force is

$$\tfrac{4}{3}\pi r^3 (\rho - \rho_1) g \tag{9}$$

where ρ is the density of the sphere and ρ_1 the density of the medium through which it falls. From Stokes' equation (equation 4) viscosity increases with increasing rate of fall of the body until a constant speed is attained when the viscous force will exactly equal the gravitational pull. Then

$$\tfrac{4}{3}\pi r^3 (\rho - \rho_1) g = 6\pi r \eta u \tag{10}$$

where u is the velocity. The distance traveled by the sphere in both solute and solvent is the same; thus, the velocity of fall is inversely proportional to the time and the relative viscosity is determined from

$$\frac{\eta_1}{\eta_2} = \frac{t_1(\rho - \rho_1)}{t_2(\rho - \rho_2)} \tag{11}$$

where ρ is the density of the sphere and t_1 and t_2 the times of fall in the two liquids of density ρ_1 and ρ_2.

Viscosity of Macromolecular Solutions

Because of the complexity of the theoretical principles involving multiple particles, explanations of viscosity concepts and calculations based on these explanations all consider the effects about an individual particle. This assumes that no particles are close enough to each other to produce or manifest any cross effects. In macromolecular solutions in which the solute molecules are much larger than the solvent molecules, a solute molecule will stretch across a number of solvent molecules' flow lines and distort the pattern of flow. Einstein[33] computed the effect produced by a spherical particle at small flow velocities and derived the equation

$$\eta = \eta_0 (1 + 2.5\phi) \tag{12}$$

where η is the viscosity of the macromolecular solute, η_0 the viscosity of the solvent and ϕ the volume fraction of the macromolecules in the total volume of the solution. This equation holds for any number of macromolecules far enough apart that overlap of the distortion of the flow lines produced by each individual macromolecule is prevented. Einstein's calculations were extended to ellipsoids by Simha.[77,104] Thus, in considering macromolecular solutions in which adjacent molecules may have an effect upon each other it is necessary that any measurements be made in solutions of high dilution and that any calculations be extrapolated to zero concentration. Solvation effects, particularly in terms of the hydrodynamic volume and in terms of the partial specific volume, must also be considered. As a result, it has become usual procedure to describe experimental data in terms of the difference between the viscosity of the solute and the viscosity of the pure solvent according to the expression

$$\eta_{spec} = \frac{\eta - \eta_0}{\eta_0} \tag{13}$$

where η_{spec} is the specific viscosity, η the viscosity of the solute and η_0 the viscosity of the pure solvent. The Einstein equation (equation 12) predicts that at infinite dilution η_{spec} is proportional to the number of macromolecules per unit volume. Therefore, the quantity η_{spec} is proportional to the concentration, c, in grams per cubic centimeter. The quantity η_{spec}/c is called the reduced viscosity and is independent of concentration at the limit of zero concentration. Extrapolation of η_{spec}/c to zero concentration gives a limiting value of η_{spec}/c, called the intrinsic viscosity, $[\eta]$, according to the relation

$$[\eta] = \lim_{c \to 0} \frac{\eta_{spec}}{c} = \lim_{c \to 0} \frac{\eta - \eta_0}{\eta_0 c} \tag{14}$$

In macromolecular solutions the reduced viscosity is concentration-dependent and is given by the Huggin's equation[52]

$$\frac{\eta_{spec}}{c} = [\eta] + k'[\eta]^2 c + \ldots \tag{15}$$

where k' is the Huggin's constant, and the Kramer equation[62]

$$\frac{\ln \eta_{rel}}{c} = [\eta] + k''[\eta]^2 c + \ldots \tag{16}$$

where η_{rel} is the relative viscosity where

$$k'' = k' - 0.5 \tag{17}$$

Crothers and Zimm[24] have recommended that at low shear stress the plot of ln η_{rel}/c versus c method of extrapolating to zero concentration is superior to the more common η_{spec}/c versus c plot. However, Ross and Scruggs[94] have shown that this is not always the case. The intrinsic viscosity, $[\eta]$, is the quantity of primary interest with macromolecular solutions and is important in that it is related to the average molecular weight of the dissolved macromolecules according to the expression

$$[\eta] = K\bar{M}^\alpha \tag{18}$$

where $\bar{M}$ is the average molecular weight and K and α are constants, the values of which depend on the nature of the polymer and solvent and on the temperature. The equation was first suggested by Staudinger[107,108] with $\alpha = 1.0$. Mark[68] and Houwink[50] first suggested α was a variable. For most polymers the value of α is less than 1. The larger the value of α, the better the solvent being used. Conversely, smaller values indicate a poorer solvent. Values of K and α can be calculated by performing two experiments on two different solutions of two macromolecules having known molecular weights. By measuring the solutions in a series of solvents at different temperatures the values of the constants for each set of conditions can be determined. Equation 18 can then be used to calculate the molecular weight for all subsequent samples of different macromolecules of the same type—for example, different DNA's. The use of this type of measurement in conjunction with sedimentation data will be discussed after the next section.

Sedimentation Analysis

Sedimentation in the ultracentrifuge is one of the most widely used methods of investigation of macropolymers, either for preparative purposes or analytically for the determination of size, shape and density, and thus molecular weight. Detailed descriptions of sedimentation methods and theories have been reviewed by Svedberg and Pedersen[111] and Schachman.[97]

The principles of preparative and analytical centrifugation are essentially the same. To determine the mass of a molecule its motion is observed under the action of known forces. For large masses the most convenient force is gravity, but for smaller particles the gravitational force is so slight that the random bombardment by the molecules of the surrounding medium negates the directing force of gravity. To overcome this difficulty the gravitational force is supplemented with force provided by a centrifuge. The macromolecular solution to be examined is placed in a centrifuge cell, which fits into a centrifuge rotor some distance from the axis of rotation. As the rotor turns, the macromolecules move through the surrounding solvent away from the axis of rotation. To determine the net gravitational force on the macromolecules corrections must be applied for frictional and buoyant forces caused by the action of the solvent on the macromolecules. If a molecule of mass m and density ρ moves through a solvent of density ρ_0 in a centrifuge cell rotating at constant angular velocity ω rads per second and the distance of the molecule from the axis of rotation is r, the linear acceleration of the molecule is $\omega^2 r$. The gravitational or centrifugal force on the molecule is $m\omega^2 r$, and the buoyant force working against the centrifugal or gravitational force is $(m/\rho)\rho_0\omega^2 r$. The net gravitational force is the centrifugal force minus the buoyant force:

$$m\omega^2 r - \frac{m}{\rho}\rho_0\omega^2 r = m\omega^2 r\left(1 - \frac{\rho_0}{\rho}\right) \tag{19}$$

The molecule thus acts as if it had an effective mass m' where

$$m' = m\left(1 - \frac{\rho_0}{\rho}\right) \tag{20}$$

This equation is difficult to use in determining the mass or molecular weight of a particle since the correct value for the density of the particle, ρ, is difficult to determine. For example, a molecule of DNA has a complex secondary and tertiary structure and is influenced by various effects due to the surrounding water molecules and other factors. To overcome this problem the reciprocal of the partial specific volume, $\bar{v}$, is used, which is the change in volume per unit mass for an infinitesimal mass:

$$m' = m(1 - \bar{v}\rho)* \tag{21}$$

When the centrifugal force is first applied the molecule accelerates very rapidly from rest, and its motion is opposed by the frictional force, fv, which increases with velocity until constant velocity is reached. At this point the frictional force is exactly balanced by the net gravitational force:

$$fv = m\omega^2 r(1 - \bar{v}\rho) \tag{22}$$

Sedimentation analysis is performed in two general ways—the sedimentation equilibrium method and the sedimentation velocity method.

Sedimentation Equilibrium Method

The sedimentation equilibrium method is used to determine the equilibrium distribution of macromolecules subjected to low net gravitational forces at low rotor speeds, about 5000 to 10,000 rpm for molecular weights in the range of 10^6 to 10^8 daltons. At the start of the experiment the macromolecules are subjected to a net gravitation force and move away from the axis of rotation. The concentration of macromolecules decreases at the top and increases at the bottom of the centrifuge cell, setting up a concentration gradient. The concentration gradient initiates a diffusion process, the molecules executing random walks. The combined processes continue until

---

* Thermodynamically the solution density ρ is the correct quantity used here.

an equilibrium is attained and no further changes in concentration occur with time; that is, the slow sedimentation is exactly counterbalanced by back diffusion. The molecules have zero velocity, and the net gravitational force is the centrifugal force minus the buoyant force. Measurement of the concentration distribution at equilibrium permits calculation of the molecular weight of the macromolecular component of the solute.

The spatial distribution of the macromolecules in the centrifuge cell is governed by the Boltzmann distribution of energy levels of the molecules. The molecules have zero velocity; thus, their kinetic energies are the same at any position in the centrifuge cell, and the only differences among the molecules at the various positions in the cell are their potential energies. The difference in potential energy between two different molecules is the work necessary to carry one molecule to a different position against the net gravitational force $m'\omega^2 r$. From the Boltzmann distribution the following equation can be derived:

$$\ln \frac{c_2}{c_1} = \frac{m'\omega^2}{2kT}(r_2^2 - r_1^2) \tag{23}$$

where c_1 and c_2 are the concentrations of macromolecules in grams per cubic centimeters at distances r_1 and r_2 from the axis of rotation and k is the Boltzmann constant. Substituting for m' from equation 21 and rearranging for m we have

$$m = \frac{2kT \ln \frac{c_2}{c_1}}{\omega^2(1 - \bar{v}\rho)(r_2^2 - r_1^2)} \tag{24}$$

Multiplying both sides by Avagadro's number, mass m becomes the molecular weight, M, and Boltzmann's constant, k, becomes R, the gas constant. Therefore

$$M = \frac{2RT \ln \frac{c_2}{c_1}}{\omega^2(1 - \bar{v}\rho)(r_2^2 - r_1^2)} \tag{25}$$

or

$$M = \frac{2RT\, d(\ln c)}{\omega^2(1 - \bar{v}\rho)\, dr^2} \tag{26}$$

This equation is independent of the shape of the molecule and the presence of bound water, which is the advantage of this method of analysis. Because the solution is not ideal, the molecular weight computed from this equation is not the true molecular weight. To overcome this problem experiments are performed with several starting concentrations, and the apparent molecular weight is obtained from a slope of a plot of log c versus

r^2. The true molecular weight is then obtained by extrapolation of the reciprocal of the apparent molecular weight versus c to infinite dilution.

In the derivation of the equations for sedimentation equilibrium it is implicit that M and $\bar{v}\rho$ are constants independent of r. This is a false assumption. The difference between M, the apparent molecular weight, and the true molecular weight is a concentration effect—since c varies with r, M must also vary with r. This is overcome by extrapolation of $1/M$ versus c to infinite dilution. Similarly, $\bar{v}\rho$ must depend on concentration and pressure, but the differences are very small and can be ignored.

Sedimentation Velocity Method

The sedimentation velocity method does not require the long time periods for establishment of equilibrium distribution as does the sedimentation equilibrium method, and is performed at centrifugal speeds of up to 100,000 rpm.

The macromolecules, which initially are distributed uniformly throughout the solution in the centrifuge cell, sediment rapidly toward the bottom of the cell. When the centrifuge cell attains constant speed, the macromolecules sediment at constant velocity, as described previously. A sedimenting boundary is formed between the supernatant at the top of the cell and the rest of the solution (the plateau region) in which the concentration of macromolecules is uniform. The movement of the boundary away from the axis of rotation corresponds to the rate of sedimentation of the macromolecules by the net forces described in equation 22.

Since the velocity, v, and the acceleration, $\omega^2 r$, are known, the ratio $v/\omega^2 r$, the velocity per unit centrifugal field, is called the sedimentation coefficient, s:

$$s = \frac{v}{\omega^2 r} \quad (27)$$

Substituting in equation 22 for s and solving for m

$$m = \frac{sf}{(1 - \bar{v}\rho)} \quad (28)$$

The frictional force, f, is difficult to determine and is dependent upon the shape of the molecule and the extent of hydration. To circumvent this difficulty the Einstein relationship between frictional coefficient and diffusion coefficient is used:

$$D = \frac{kT}{f} \quad (29)$$

where D is the diffusion constant. Thus,

$$m = \frac{kTs}{D(1 - \bar{v}\rho)} \quad (30)$$

Multiplying both sides by Avagadro's number, the Svedberg equation is obtained:

$$M = \frac{RTs}{D(1 - \bar{v}\rho)} \qquad (31)$$

The unit of the sedimentation coefficient, s, is seconds but is reported in Svedbergs, S, in honor of the pioneer of centrifugation, where $S = 10^{-13}$ seconds. Experimentally it is the sedimentation coefficient that is determined according to equation 27. The velocity, v, is the change of distance of the boundary from the axis of rotation with time:

$$v = \frac{dr}{dt} \qquad (32)$$

where r is the distance from the axis of rotation. Equation 27 then becomes

$$s = \frac{1}{\omega^2 r} \cdot \frac{dr}{dt} = \frac{1}{\omega^2} \cdot \frac{d(\ln r)}{dt} \qquad (33)$$

and s is determined from the slope of a plot of log r versus time. The value of s is dependent on the solvent, temperature and macromolecular concentration. The concentration dependence is seen in equation 31, since the frictional coefficient varies with concentration and D is a function of f. To circumvent these factors experiments are performed at progressively decreasing macromolecular concentrations in a solvent having an electrolyte concentration sufficient to depress undesirable charge effects. The values of s obtained are corrected to a reference solvent of viscosity and density of water at 20°C and then extrapolated to infinite dilution. The extrapolated value of the sedimentation coefficient is represented by S^0_{20w} and is independent of solute-solute interactions. It is used, together with the extrapolated corrected diffusion coefficient, in the Svedberg equation (equation 31) to calculate the anhydrous molecular weight of the macromolecule.

The Svedberg equation for molecular weight of a molecule is more complex than the equation for the sedimentation equilibrium method, since the diffusion constant as well as the partial specific volume of the molecules being examined must also be known. The measurement of accurate diffusion coefficients of macromolecules has recently been aided by the development of the technique of optical mixing spectroscopy.[17,30,31] A laser light source of high spectral purity is used as the source of incident light. The spectrum of the light scattered from a solution of macromolecules is measured with an ultra-high resolution self-beating optical mixing spectrometer. The spectral width is directly proportional to the diffusion coefficient of the macromolecules scattering the light. Measurements are accurate to about 1 per cent and take approximately one hour to complete, compared with days necessary for conventional methods. The combined application of the equilibrium and velocity sedimentation methods gives information on the size and the relation between the shape and extent of hydration of a macromolecule in a particular solvent.

The Archibald Method of Sedimentation Analysis

The basic equations for sedimentation equilibrium apply to two points in a sedimentation cell. The meniscus and the bottom of the cell can be chosen as these points and are unique in that no net flow occurs through these points. Archibald[1] pointed out that this applies at any time during the approach to equilibrium in the cell and that this can be used to determine molecular weights if the distribution of concentration in these regions is known. The concentration distributions can be calculated by considering the amount of macromolecular solute crossing a surface of area a at position r in a cell owing to diffusion and sedimentation during a time interval, dt. The diffusion component is expressed by Fick's first law of diffusion, which relates the transport of material to the concentration gradient by the equation

$$-Dar\left(\frac{\partial c}{\partial r}\right) dt \qquad (34)$$

where $\partial c/\partial r$ is the concentration gradient. The sedimentation component is the product of concentration, area and rate of sedimentation:

$$car(s\omega^2 r)\, dt \qquad (35)$$

The combined relation of equations 34 and 35 is

$$ar\left[cs\omega^2 r - D\left(\frac{\partial c}{\partial r}\right)\right] dt \qquad (36)$$

At the meniscus and at the bottom of the cell, since there is no net flow, at these points the expression in the brackets is zero and

$$c_m s\omega^2 r_m = D\left(\frac{\partial c}{\partial r}\right)_m \qquad (37)$$

for the meniscus and

$$c_b s\omega^2 r_b = D\left(\frac{\partial c}{\partial r}\right)_b \qquad (38)$$

for the bottom. Solving for s/D in the Svedberg equation (equation 31) gives expressions for molecular weight at the meniscus and at the bottom of the cell:

$$M_m = \frac{RT}{(1-\bar{v}\rho)\omega^2} \frac{\left(\frac{\partial c}{\partial r}\right)_m}{r_m c_m} \qquad (39)$$

and

$$M_b = \frac{RT}{(1-\bar{v}\rho)\omega^2} \frac{\left(\frac{\partial c}{\partial r}\right)_b}{r_b c_b} \qquad (40)$$

Application of Archibald's method is made at the start of sedimentation equilibrium and is rapid and accurate, the molecular weights obtained having the same validity as those determined at equilibrium. Values for $\partial c/\partial r$ and c are readily obtained from refractive index measurements using schlieren optics.[97] If the macromolecular solute is homogeneous, $M_m = M_b$; if not, the method also detects heterogeneity. The method has become one of the most widely used for determination of molecular weights of both large and small molecules.

Density Gradient Centrifugation

The large centrifugal forces which can be generated in modern centrifuges have been used to advantage in a sedimentation equilibrium method in which macromolecules are centrifuged in a density gradient.[79] A density gradient is obtained by centrifugation of a solution of heavy salt, such as cesium chloride, until equilibrium is reached, as defined by equation 23. Since there is a concentration gradient between the top and the bottom of the centrifuge cell, there is a density gradient. When a macromolecule whose density falls between the two extremes of density of the solution is also centrifuged in the same solution, it will accumulate at its lowest energy position, which corresponds to the position where its effective mass, m', is zero (see equation 20). At this point the density of the solvent will equal the density of the macromolecule. The distribution about this point is narrow because of the high centrifugal speeds used to set up the density gradient, and the resolution that can be obtained by this method is excellent. Meselson and Stahl[78] in a classic experiment, separated ^{15}N-containing E. coli DNA from normal ^{14}N E. coli DNA; they were able to distinguish between these two types of molecules and also molecules of E. coli DNA containing half ^{15}N and half ^{14}N DNA, which banded at a density in between that of the ^{15}N and ^{14}N molecules. The method has also been used to determine the base composition of DNA, since the buoyant density of DNA in CsCl is directly proportional to its guanine-cytosine (G-C) content.[69,92,110] In a study of many different DNA samples Schildkraut, Marmur and Doty[101] obtained the relation

$$\rho = 1.660 + 0.098\,(GC) \qquad (41)$$

where ρ is the buoyant density in grams per cubic centimeters and GC the mole fraction of guanine plus cytosine. Cesium salt density gradient sedimentation has developed into one of the most widely used techniques for purification and analysis of nucleic acids. The technique has been the subject of a recent review.[112]

Band Sedimentation

Vinograd, Bruner, Kent and Weigle[124] have introduced a modification for sedimentation velocity experiments in which a thin lamella of a macromolecular solution is layered onto a denser liquid of electrolyte solvent. The migration and spreading of the macromolecular band is then followed as a function of time, macromolecules with the same sedimentation coefficient sedimenting through the liquid in a narrow band, and a mixture of non-interacting macromolecules resulting in a number of bands. One of the advantages of the method is that the small density gradient necessary for stability in the leading part of the band is generated automatically by diffusion of small molecules between the thin lamella of solution first introduced and the main liquid column in the centrifuge cell. Also, the moving macromolecules are always contained within bands or zones, compared with the uniform solutions and boundaries in the conventional sedimentation velocity procedure. Vinograd has pointed out that the method is analogous to the zone-velocity centrifugation method used in the preparative ultracentrifuge, but differs from it in that the stabilizing density gradient is not formed throughout the bulk of the solution prior to the application of the sample and only involves shallow density gradients.[124] Thus, the conditions for ideal sedimentation velocity experiments can be approached. Vinograd has introduced the term band centrifugation to describe this method. The method has been applied to a wide variety of macromolecular species, including DNA, nucleolar RNA, polyribosomes and viruses. In a series of papers Vinograd and co-workers[122-124] have described the theory of the method in detail and have shown that the method can be applied to the analysis of homogeneous, non-interacting molecules and for measurement of both sedimentation and diffusion coefficients.

Zone-Velocity Sedimentation in Preformed Density Gradients

This is a preparative technique for separation of macromolecules according to their sedimentation coefficients.[28,73] A small volume of the macromolecular solution is layered on top of a continuous density gradient. The gradient is usually a sucrose solution and is formed in the centrifuge cell from dilute and concentrated solutions with the aid of a mechanical mixing and delivery device. The centrifuge cells are cylindrical tubes and are spun in a swinging bucket rotor. Large zonal centrifuge rotors, in which the rotor is in fact the centrifuge cell, are also used for large volume or continuous flow preparations. The macromolecules sediment in a narrow zone through the gradient, which stabilizes the system against convection. Mixtures of molecules are separated

116 / METHODOLOGY AND PROPERTIES

into a number of discrete zones. Rotor speed and duration of centrifugation are selected to allow maximum separation without pelleting. The gradient is fractionated by one of several methods of removing the centrifuge contents, analyzed and sedimentation patterns plotted of analysis versus direction and distance of sedimentation. Inclusion of a standard macromolecular compound of known sedimentation coefficient in the macromolecular solution allows approximation of the coefficients of the other components of the solution.

The Analytical Ultracentrifuge

Sedimentation equilibrium and velocity studies are performed in analytical ultracentrifuges. Commercially available models have speed capacities up to 100,000 revolutions per minute. The rotor is spun in an evacuated chamber to minimize heating effects, and the cell has a sector shape, shown in Figure 4-3, so that the sedimenting particles travel along the radii, eliminating other components of force on the particles other than gravity and friction. The cells are fitted with quartz windows to permit observation of the sample during sedimentation. The position of the sedimenting boundary may be followed by noting changes in absorption of light at a predetermined wavelength, depending on the sample being used—for example, with nucleic acids this would be in the ultraviolet region—or by determining differences in the refractive index of the particles from that of the solvent. The majority of commercially available centrifuges are fitted with schlieran optics; these are based on the refractive index principle, in which the refractive index of the component being measured is proportional to the solute

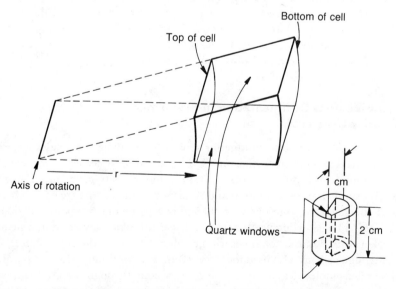

Figure 4-3. Schematic diagram of an ultracentrifuge cell.

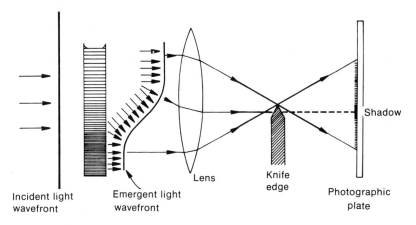

Arrows indicate directions of normals to the wavefront.

Figure 4–4. Schematic diagram of a schlieren optical system.

concentration. Incident light falling on a cell with a concentration gradient results in an emergent wavefront, as depicted in Figure 4–4. In the curved part of the line the emergent wavefronts of the light will be directed away from the normal, whereas the wavefronts in the areas of pure solute and pure solvent will be parallel and in the same plane as the incident light. If a lens is placed between the cell and a photographic plate, emergent wavefronts whose normals have not been deviated will pass through the focal point of the lens and form inverted images on the photographic plate. Emergent light which has deviated from the normal would pass below the focal point. By placing an opaque knife edge just below the focal point all these deviated rays are cut out and no light from the concentration gradient range appears on the screen (Figure 4–4). In this manner regions of rapid changes in refractive index appear as shadows on the screen, giving the name schlieran, or streaked, optics to the system. As sedimentation proceeds the boundary moves down and the shadow up. Other more complicated systems are available which translate the deviations of the wavefront normals into horizontal reflections and convert vertical change in wavefront direction into a horizontal deflection. This gives a curve whose deflection is proportional to the original wavefront deflection from the horizontal, which is proportional to the refractive index gradient. Thus, the area under the curve is proportional to the concentration. By use of such optical systems the rate of change of the solute with distance can be observed.

Relationship Between Sedimentation and Viscosity

One of the advantages of the sedimentation velocity method is that it not only gives a measure of the molecular weight but also provides a measure of the frictional coefficient. Mandelkern and Flory[67] and Scheraga and Mandelkern[100] combined the relationship of the sedimentation coefficient to the frictional coefficient and the viscosity to the frictional coefficient in the equation

$$M^{2/3} = \frac{S^0[\eta]^{1/3}\eta_0 N}{10^{13}\beta(1 - \bar{v}\rho)} \quad (42)$$

where S^0 is the sedimentation coefficient extrapolated to zero concentration, $[\eta]$ the intrinsic viscosity of a polymer in a solvent of viscosity η_0 and solution density ρ, $\bar{v}$ the partial specific volume of the polymer, N Avogadro's number and β an empirical coefficient which depends on the form of the molecules. For random coils, such as high molecular weight DNA, the value of β is practically constant.[35] This relationship was re-examined by Crothers and Zimm,[24] Aten and Cohen[2] and Eigner and Doty[32] for high molecular weight DNA. They showed that by working under careful experimental conditions of low shear and low rotor speeds it was possible to obtain good values for the intrinsic viscosity and sedimentation coefficients of DNA molecules.

Eigner and Doty[32] determined the sedimentation coefficients and intrinsic viscosities of a number of DNA samples and selected other values from the literature. In plots of sedimentation coefficient versus intrinsic viscosity they showed that most DNA molecules are members of a homologous series in a relationship which holds over a thousand-fold range of molecular weight. The importance of the plot is that it is independent of molecular weight measurements. From other analyses of their own and from previous data from other laboratories they showed also that light scattering determinations of molecular weight are reliable only below 3,000,000 daltons and that the value of the Mandelkern-Scheraga-Flory constant, β, remains very close to 2.4×10^6 throughout the molecular weight range of 200,000 to 200,000,000 daltons. Using this value for β, S^0 and intrinsic viscosity values can be converted easily to molecular weights.

Crothers and Zimm[24] presented a method for empirical correction for excluded volume effects of DNA by plotting the results of viscosity and molecular weight analyses on a log-log plot (log s versus log M) over the molecular weight range of 2.0×10^5 to 1.3×10^8 daltons. Hearst and Stockmayer[45] had previously described an equation relating the sedimentation coefficient and the molecular weight for ideal worm-like chain molecules:

$$S = b + KM^{1/2} \quad (43)$$

The equation is a linear function of the square root of the molecular weight, and low molecular weight data (below 10^7 daltons) for DNA fitted the relation.[45] Zimm and Crothers pointed out that this function neglected excluded volume effects. By representing S as a linear function of M^α, they described the equation

$$s - b_s = K_s M^\alpha \quad (44)$$

where α, K_s and b_s are constants. Plotting $s - b_s$ against M on logarithmic scales gave a straight line of slope α. By selecting 2.7 as the constant b_s, the same value used by Hearst and Stockmayer,[45] the exponent α was 0.445. In a similar treatment of viscosity data they described the equation

$$[\eta] + b_\eta = K_\eta M^\alpha \qquad (45)$$

In plots of $[\eta] + b_\eta$ versus M on logarithmic scales they selected a value for b_η of 5 decilitres per gram to give the best straight line. With this value the slope of the plot gave α a value of 0.665. Substituting the values from equations 44 and 45 into the Mandelkern-Scheraga-Flory equation gives a value for β:

$$M^{2/3} = \frac{(S - 2.7)([\eta] + 5)^{1/3}\eta_0 N}{2.27 \times 10^{19}(1 - \bar{v}\rho)} \qquad (46)$$

The value of 2.27×10^6 was in good agreement with the value obtained by Eigner and Doty[32] of 2.4×10^6 in the limit of high molecular weight. Rosenberg and Studier,[93] in a study of T7 DNA and the relationship of sedimentation, viscosity and ionic strength of the solvent, showed that the Mandelkern-Scheraga-Flory relationship held over the entire range of ionic conditions between 0.0013 M and 1 M sodium ion concentration. They observed a decrease in values of β for native T7 DNA at higher ionic strengths, and similarly for alkaline denatured single-stranded DNA, which correlated with a decrease in the intrinsic viscosity of native DNA by $\frac{1}{3}$ between 0.02 M and 1 M NaCl. However, this was not paralleled by a change in the S^0_{20W} values. They suggested that the discrepancy between the effects of ionic strength on $[\eta]$ and S^0_{20W} was due to a change in $\bar{v}$ and cited a previous report by Cohen and Eisenberg[18] on native calf thymus DNA in which it was shown that $\bar{v}$ increases by approximately 5 per cent between 0.2 M and 1 M NaCl. Previously workers had assumed that $\bar{v}$ does not change with ionic strength. Rosenberg and Studier,[93] using Cohen and Eisenberg's values for $\bar{v}$, recalculated a value for β. The new value was 2.21×10^6, which compared favorably to the value of 2.18×10^6 in 1 M NaCl, indicating no change at higher ionic strength. Thus, the change in $\bar{v}$ accounts for the insensitivity of S^0_{20W} to changes in the expansion of native DNA at high ionic strength. On this basis they suggested that intrinsic viscosity is a more sensitive and reliable measure of slight conformational changes in native DNA than S^0_{20W} values. In a series of parallel studies they also came to the conclusion that S^0_{20W} is a better index of conformational changes in single-stranded DNA since the intrinsic viscosity becomes too small at high ionic strengths to measure conveniently. Opschoor, Pouwels, Knijnenburg and Aten[84] have extended studies of the intrinsic viscosity and sedimentation coefficients of linear DNA molecules to circular, double-stranded DNA's. Bloomfield and Zimm[6] had previously calculated the ratio between the intrinsic viscosities and sedimentation coefficients of circular and linear molecules in two limiting cases and had correlated these as a function of an expansion parameter, ε. Opschoor, et al., showed that the equations relating intrinsic viscosity and sedimentation coefficients as a function of molecular weight, as first developed by Crothers and Zimm for linear DNA's, could be written for circular, double-stranded DNA. In solvents of ionic strength 0.2 M the equations are

$$[\eta] = 8.64 \times 10^{-4} M^{0.665} - 3.9 \tag{47}$$

and

$$s = 1.76 \times 10^{-2} M^{0.445} + 2.9 \tag{48}$$

Experiments on double-stranded ϕX174 replicative form DNA showed that both the intrinsic viscosity and sedimentation coefficients agreed well with the theoretical values predicted from equations 47 and 48. Values for b_s and α in equation 44 have recently been recalculated by Freifelder,[38] who used very accurate values for molecular weights of T4, T5 and T7 DNA's.[3,30,64] Plotting $\log(s - b_s)$ versus $\log M$ for various values of b_s, the best curve obtained was for $b_s = 2.8$ for which $\alpha = 0.479$. This gave a value for K_s of 0.834×10^{-2}.

Determination of Molecular Weights by Electron Microscopy

This is obviously not a hydrodynamic property but is included in this section as an addendum to the various techniques for determining molecular weights of nucleic acids.

Electron microscopy affords a relatively simple method of determining molecular weights of nucleic acid molecules. The technique became important after Kleinschmidt described the monolayer spreading technique.[61] The basic technique, which has subsequently been modified many times by numerous workers, involves the spreading of a monolayer of protein, usually cytochrome c, in which the nucleic acid molecules are suspended on a solution in a Petri dish. A droplet of the protein solution is introduced onto the liquid in the Petri dish. As it spreads over the surface of the liquid a monolayer is formed, and the more electron dense cytochrome c molecules cluster around the nucleic acid molecules that are held in the monolayer. Samples of the monolayer are then caught up on electron microscope grids, dried and examined under the electron microscope. A large number of nucleic acid molecules can be examined with this technique, and if the original nucleic acid preparation is homogeneous and if the molecules have not been broken by shear during preparation, their lengths can be measured from photographs, either with a map measuring tool or by laying a silk thread along the photographic image and then measuring the length of silk thread. From the density per unit length of the nucleic acid the molecular weight can be computed. The length of a molecule (l) is related to the molecular weight (M) by the equation

$$M = M'/l \qquad (49)$$

where M' is the molar linear density. M' has been calculated by the majority of workers from the Watson-Crick β configuration for double-stranded DNA. It is assumed that this configuration occurs during the preparation of the sample for electron microscopy. Lang[64] has recently made the calculation by calibration with DNA's of independently determined molecular weights and obtained a value of 2.07×10^{10} daltons per centimeter for double-stranded DNA with common bases and average base composition. Care must be exercised during this procedure to ensure that the correct magnification of the sample is recorded and that the DNA samples are maintained at the correct ionic strength, since length depends on the ionic strength and composition of the medium supporting the protein film.[54,65] Since the method is a visual one, a main advantage over other systems is that measurement of poor, sheared preparations of molecules is avoided. A sufficient number of individual molecules can be measured to obtain a proper statistical evaluation of the results. However, as with all other methods of measurement of molecular weight this method should not be relied upon alone, but should always be correlated with other methods, such as sedimentation and intrinsic viscosity measurements.[38] Apart from the obvious application for measurement of molecular weight, the method has also been used for visualization of deletion mutants[27,128] (by comparing normal and deletion mutant strains of various bacteriophages) and visualization of RNA synthesis in eukaryotic nuclei and bacteria,[82] and has also been used in the development of structural comparisons (by examination of various molecules under various conditions of denaturation and the construction of denaturation maps).[53,55] This last application was pioneered in Inman's laboratory and allows correlation with genetic and chemical evidence of structural parts of DNA molecules. Deletion mutations have been particularly interesting in electron microscopic visualization because hybridization of one DNA strand of a deletion mutant with the complementary strand of the wild-type (undeleted) molecule results in the looping out of that portion of the wild-type molecule which is deleted, and this loop can be seen and measured in the electron microscope.[128] In this way the true extent of the deletion mutant can be measured directly. This also confirms that all the other parts of the molecule are complementary, that it is a true deletion in the middle of the molecule and also confirms the hybridization methodology used. The basic principle of hybridization will be discussed in the section on thermal denaturation.

Flow Birefringence

This technique has not been investigated extensively on nucleic acids,[7,83,85] although it provides the potential of measuring the length of any long, thin molecule and should be applicable to DNA.[85] Flow birefringence depends upon two basic properties. First, the molecules to be measured must be optically anisotropic; that is, their optical properties must depend on their orientation relative to the plane of polarization of the light used to examine them. When a molecule has differences in polarizability along different directions, which leads to corresponding differences in refractive index for the light that is polarized, the term birefringence (double refraction) is used. The second property, rotary diffusion, is based on the rotary motion of a molecule in solution. The rotary motion of a molecule can be considered to be in one dimension if the motion of every point of the molecule is assumed to be a simple rotation about a single axis drawn in space which passes through the center of gravity of the molecule. The motion is described by a single angle, ϕ, between a reference axis of the molecule and a reference axis in space, both axes passing through the center of gravity at right angles to the rotation axis. In a dilute solution of molecules, the rotation of all the molecules will be one-dimensional if the particles are oriented in a single plane—for example, by flow—and the rotational coordinate can be represented by a single parameter, ϕ, for all molecules (Figure 4–5). The density in ϕ-space, $\rho(\phi)$, is used to describe the number of molecules per cubic centimeter of solution with orientation between ϕ and $\phi + d\phi$, which is $\rho(\phi)\, d\phi$. In the situation in which the molecules are subjected to random collisions with solvent molecules only, $\rho(\phi)$ will be a constant independent of ϕ, since all values of ϕ are equally probable. If a torque is applied to the system, the molecules will become oriented in solution; some orientations

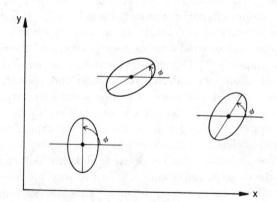

Figure 4–5. One-dimensional rotation. Each ellipse corresponds to a reference plane (parallel to the xy plane) of a particular particle. Rotation occurs about the z axis which is perpendicular to the xy plane and passes through the center of gravity of each particle. The angle ϕ is the coordinate describing the rotational position.

become more probable than others, and $\rho(\phi)$ is dependent on ϕ. If the torque is removed, a gradual redistribution of orientations occurs until $\rho(\phi)$ is again constant. This process is rotary diffusion, and $\rho(\phi)$ is a variable of rotary motion.

The theoretical treatment of rotary diffusion is similar to translational diffusion and analogous to Fick's first law. The relation of the movement of the number of molecules through the angle ϕ in a direction of positive ϕ in time dt given by

$$-\Theta\left[\frac{\partial \rho(\phi)}{\partial \phi}\right] dt \qquad (50)$$

where Θ is the rotary diffusion coefficient with units of sec^{-1}. The rotary diffusion coefficient depends on molecular form and dimensions, as does the linear diffusion coefficient. Perrin[87] has shown that if a long thin molecule is thought of as a prolate ellipsoid, the rotary diffusion coefficient for rotation about the b axis is given by

$$\Theta_b = \frac{3kT}{16\pi\eta a^3}\left(2\ln\frac{2a}{b} - 1\right) \qquad (51)$$

where a is the long axis of the ellipsoid and b the axis of rotation of the ellipsoid. The small axis, b, appears only in the logarithm and a large change in b will give only a small change in Θ_b. Also, Θ_b is approximately inversely proportional to $1/a^3$. Thus, the rotary diffusion coefficient may be used to measure the length of a molecule.

To measure the rotary diffusion coefficient experimentally flow birefringence is used, which utilizes the shearing force produced in a flowing liquid to orient the asymmetric molecules. The orientation is opposed by rotary diffusion, and the dependence of the extent of orientation on the shearing force is a measure of the rotary diffusion coefficient. The apparatus consists of two concentric cylinders, somewhat similar to the Couette viscometer, one of which rotates while the other is fixed. The space between the cylinders contains the solution to be studied and, as in the viscometer, the rotation of one cylinder sets the solution in motion. Thus, a velocity gradient is set up in the fluid, the layer adjacent to the moving cylinder moving the fastest with a decreasing order of velocity toward the layer adjacent to the stationary cylinder. Provided that the difference in the diameters of the cylinders is small, the velocity gradient is close to constant. The solution to be analyzed is viewed between a crossed polarizer and Nicol prism analyzer, the analyzer being at right angles to the polarizing prism. Because of the velocity gradient and rotary diffusion the molecules make a most probable angle χ with the streamlines (see Figure 4–6). When the solution is viewed in the polarized light at four positions, the optical axis of the particles will be parallel or perpendicular to the plane of polarization Light passing through the solution at these positions will travel as a single polarized beam, will not

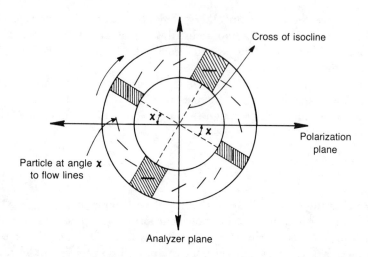

Figure 4-6. Schematic view looking down on an apparatus of the experimental manifestation of flow birefringence. The orientation of molecules is shown by the rods in the solution. Extinction occurs only when the axes of the molecules are parallel to either the polarizer or analyzer directions.

be split and will be completely stopped by the analyzer prism. These four dark areas are seen in the emerging light at the corners of a cross called the cross of isocline. The cross of isocline makes the angle χ with the cross formed by the planes of polarization of the two Nicol prisms and is called the extinction angle. The extinction angle, χ, is identical with the orientation angle, ϕ, so ϕ can be measured directly. Since the distribution of molecules depends on the velocity gradient and the rotary diffusion coefficient, and the velocity gradient is known, measurement of the extinction angle permits calculation of the rotary diffusion coefficient. By taking measurements at different temperatures and viscosities it is possible to determine whether the particle is rigid. For example, if a particle were a random coil, its length would increase with increasing velocity gradient and the rotary diffusion coefficient would decrease with increasing gradient. By use of this technique tobacco mosaic virus has been shown to act like a rigid rod and to have a length of 3400 Å.[7,83] This compares favorably to a measurement of 3000 Å obtained in the electron microscopic studies. Flow birefringence has been used in the study of numerous protein molecules and may have more applicability to DNA solutions and RNA solutions than has been fully exploited.

SPECTROSCOPIC PROPERTIES

All molecules possess a series of discrete quantized energy states. These energy states differ from each other in the energies of the electrons in the electrostatic field of the atomic nuclei, in the vibrational energies of these nuclei relative to one another, and in the rotational energy of the molecule as a whole. Additional energy levels are created if a molecule is placed in an external electric or magnetic field, since these fields may interact with the rotation of the molecule or with the spin of the electrons and nuclei. The internal energy of a molecule does not vary continuously, but if a correct amount of external energy is supplied to a molecule for transition from one energy state to another, the supplied energy may be absorbed and the molecule may undergo the corresponding transition, provided that the transition probability is favorable. Study of such transitions provides information on the structure and organization of molecules. The external energy sources used are electromagnetic radiation—either infrared, visible, ultraviolet or X-rays—thermal energy, electric fields, magnetic fields, charged particles (protons and alpha particles) and uncharged particles (neutrons). Some of these energy sources have been used extensively in investigations of the structure and molecular properties of nucleic acids.

Electromagnetic radiation is quantized and from the standpoint of energy transfer behaves as though it were composed of discrete packets of energy, which are called photons. The energy, E, of each photon is related to the frequency of the radiation, ν, by Planck's constant, h (6.624×10^{-27} erg-sec):

$$E = h\nu \tag{52}$$

The frequency is the product of the velocity of light, c (3×10^{10} cm/sec) and the reciprocal of the wavelength, λ, or wave number, $\bar{\nu}$, which gives the relative photon energies:

$$\nu = \frac{c}{\lambda} = c\bar{\nu} \tag{53}$$

Therefore, the energy per photon or quantum is

$$E = h\nu = \frac{hc}{\lambda} = hc\bar{\nu} \tag{54}$$

Thus, transitions of the three energy states in molecules—electronic, vibrational and rotational—and of the energy states induced by electric and magnetic fields will be dependent on the wavelength of the radiation source. In most small molecules the separation between electronic energy levels is approximately 10^{-11} ergs or larger. Radiation in the ultraviolet range, where λ is approximately 2000 Å, provides a source of photons of the correct order of magnitude of energy to induce typical electronic transitions. The separation between vibrational energy levels is of the order of 10^{-13} to 10^{-12} ergs. Infrared radiation, where λ is 1 to 20 microns, provides photons which can induce typical vibrational transitions. The separation between rotational energy levels is of the order of 10^{-16} ergs. Radiation frequencies in the radio range, where λ is 1 millimeter to 1000 meters, provide photons which can induce rotational transitions and transitions between energy levels in electric and magnetic fields. Molecules will therefore absorb characteristic frequencies corresponding to the exact separations between their energy levels. It is this physical property of molecules that has made spectroscopy so widely used and applied in all fields of physical chemistry.

Ultraviolet and Visible Spectra

Ultraviolet and visible spectra of compounds are a measure of the energy involved in electronic transitions. The spectrum of a compound taken when the compound is in the gaseous state, such as with benzene, consists of a series of sharp peaks extending over a broad wavelength range of as much as 1000 Å. The series of peaks is due partly to the existence of several closely related excited states, and partly to the fact that electronic transitions may occur with or without simultaneous changes in vibrational energy. This fine structure of the spectrum is observed because in the gaseous state the individual molecules absorb light without interference by neighboring molecules. However, electronic spectra are extremely sensitive to inter-molecular interaction, and in the spectrum of a compound in solution the fine structure is washed out by the variability in the energy of interaction with neighboring molecules. Thus, ultraviolet and visible spectra of compounds in the liquid state usually consist of broad bands.

The broad spectra of nucleic acids can be broken down into some fine structures at 283°K.[105] This has not been investigated further, but it is possible that studies in this area for the investigation of optical properties of nucleic acids, particularly oligonucleotides, might prove fruitful. Molecules with the most intense observable electronic spectra are those with conjugated double bond systems, absorbing ultraviolet light below 3000 Å. Nucleic acids fall into this group of compounds. The broad absorption bands extend to the shortest wavelengths experimentally obtainable, approximately 1800 Å, limited because most substances, including air and quartz, begin to absorb near 2000 Å. Electronic spectra are equally sensitive to intra-molecular changes and inter-molecular interactions. Small changes in structure lead to pronounced differences in spectra; for example, at pH 1, pH 7 and pH 14 the compound cytosine has three different spectra owing to two dissociable hydrogen ions (Figure 4–7). This property has made the study of electronic spectra particularly important. However, because of the extreme sensitivity of spectra to intra-molecular changes, theoretical interpretation and chemical explanation of the changes is extremely difficult. Compared to infrared spectra, electronic spectra give very little direct information concerning the structure of macromolecules. Ultraviolet and visible spectra of nucleic acids have been used widely to measure concentration, to follow rates and equilibria of reactions, to follow structural changes induced by other factors and for identification of the individual base components of the

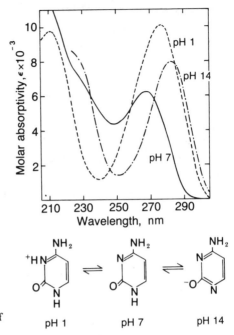

Figure 4–7. Absorption spectra of cytosine at pH's 1, 7, and 14.

macromolecules after hydrolysis to the constituent components. They have been used also in attempts to characterize oligonucleotides of different sequence. Ratios of the extinction at various wavelengths of oligoribonucleotides of known sequence have been written into computer programs to provide an information catalogue for sequence analysis of unknown oligonucleotides.[116] The sum of the spectra of the purine and pyrimidine components of a nucleic acid, in the correct molar proportions, totals the spectrum of the polymerized macromolecule with respect to wavelength characteristics. However, the intensity of absorption of the polymer (nucleic acid) is much less than the sum of the intensities of absorption of the individual nucleotides of which the macromolecule is composed. In DNA preparations the intensity is approximately 40 per cent less than observed in a mixture of the corresponding nucleotides.[4] This effect is known as hypochromicity and is due to interaction between the absorbing purine and pyrimidine bases when these are formed into their regularly ordered, base-paired configuration in the double-stranded helical molecule. The interactions involve hydrogen bonding between the opposing bases in the base pairs themselves and other interactions between the adjacent bases on the same strand, the so-called stacking forces.[115] Measurements of changes in hyperchromicity, the reverse effect, increase in absorbance owing to reduction of the interactions by externally applied forces, have become a very important tool in the measurement of changes in ordered structure in nucleic acids by various denaturing agents.

Thermal Denaturation of Nucleic Acids

In 1959, Marmur and Doty[69] showed that when DNA in solution was subjected to changes in temperature, a plot of the change in absorbance or hyperchromicity at 260 nm versus temperature gave a transition curve that was typical for each type of DNA studied and could be related to the molar proportions of the four constituent nucleic acid bases in the DNA. The midpoint of the transition curve, the T_m, when plotted for various DNA samples versus the percentage GC content of the DNA, resulted in a straight line relationship, showing that the transition temperature, T_m, was directly proportional to the base composition of the DNA.[71] The relationship is described by the equation

$$T_m = 69.3 + 0.41(\text{GC}) \tag{55}$$

where T_m is in degrees centigrade and GC is the mole percentage of guanine plus cytosine. Thermal denaturation can also be followed with the aid of many other parameters, such as light scattering, optical rotatory dispersion, density gradient centrifugation or transformation,[72] but hyperchromicity has been most widely used for measuring this phenomenon. In associated studies Marmur and Doty showed that once the DNA was denatured, denaturation being a separation of the two complementary strands of the DNA molecules, the DNA molecules could be reassociated, or renatured by slow cooling.[70]

The phenomenon of renaturation was the impetus for later studies on the hybridization technique for nucleic acids.[39,40,43] Thermal transitions of nucleic acids have been studied widely in DNA, RNA and on synthetic polynucleotides. They have been used, as previously mentioned, for determination of DNA base compositions, in investigation of other effects on the DNA structure, such as salt effects,[42,66] in investigation of the kinetics of the denaturation with temperature[74,114,129] and they have ultimately resulted in studies on partial denaturation,[41] complemented by associated studies with the electron microscope in Inman's laboratory.[53,55] From these studies certain basic information has become generally accepted in nucleic acid chemistry. Heat disrupts the hydrogen bond formation between the base pairs of the DNA double-helix, resulting in separation of the strands into two single-stranded entities. Most of the hyperchromicity observed in nucleic acids, both RNA and DNA, is due to these base pair relationships, with the hydrogen bonds restricting the double bond resonance of the nucleic acid bases. However, other optical studies, which will be described later, have shown that an appreciable amount of hyperchromicity is due to stacking forces between the adjacent nucleotides. The steepness of the melting curve and the midpoint of the transition curve have been subjected to extensive theoretical treatment[22,23] and have become standard accepted parameters for evidence for double-strandedness, or secondary structure, of the hydrogen bond-type in nucleic acids. Recently, thermal transitions have been shown to take place in small pieces of nucleic acid at temperatures

SPECTROSCOPIC PROPERTIES / 129

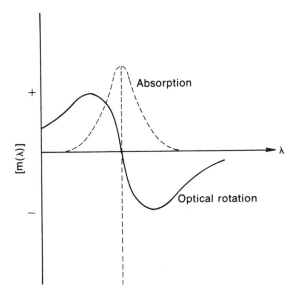

Figure 4-8. Schematic of the absorption spectrum and ORD spectrum of a hypothetical substance.

lower than ambient,[8,36,98,99] suggesting that stacking forces can also be disturbed by thermal effects and can be measured through differences in electronic transitions.[26]

Optical Rotatory Dispersion and Circular Dichroism

Optical rotatory dispersion and circular dichroism are based on electronic transitions in the ultraviolet region of the spectrum, and have been used in the study of the molecular structure of nucleic acids. Both methods are dependent on the asymmetry of the atoms within the nucleic acid molecules, which gives rise to rotation of the plane of polarized incident light.

Optical rotatory dispersion is a dispersive phenomenon and is the measure of the optical rotation and its variation with wavelength in polarized light. The variation with wavelength is shown in Figure 4-8. The negative Cotton effect involves two extrema and an inflection point. Rotation persists at wavelengths far from the Cotton effect. Analysis of optical rotatory dispersion spectra involves the identification of the optically active chromophores in the molecule and knowledge of the frequency shapes and amplitudes of the bands in the spectra. A compound, such as a nucleic acid, with a number of absorption bands will exhibit a number of Cotton effects, complicating the spectrum and its analysis. Optical rotatory dispersion has been used extensively in the study of the helical conformation of proteins, but in nucleic acids the studies have centered mainly on the optical properties of oligonucleotides, particularly dinucleotides and trinucleotides.[15,57-59,126,127] These studies have provided some insight into the neighbor-neighbor interactions and stacking force relationships in these types of compounds. Studies on synthetic nucleic acids, RNA and DNA,[96,120] have not been as rewarding in terms of conformation analysis as similar studies on proteins. Optical rotatory dispersion data on nucleic acids does not provide much more information than ultraviolet absorption data.

Circular dichroism depends on polarized light, as does optical rotatory dispersion, but the plane polarized light is resolved into two circularly polarized components by passage of the monochromatic polarized beam through an appropriate quarter waveplate. When the light passes through the optically active asymmetric molecule under study, one of the circularly polarized components will be absorbed to a greater extent than the other, resulting in light that is elliptically polarized. The difference in optical density of the two circularly polarized components is measured directly, then converted to a difference in extinction coefficient. Circular dichroic spectra are very similar to absorption spectra (Figure 4–9) if the left circularly polarized light is absorbed to a greater extent than the right, giving a positive ellipticity. If, however, the reverse is true, then a negative ellipticity band will occur.

As with optical rotatory dispersion, a number of studies on nucleic acids by circular dichroism spectra have been performed.[5] In oligonucleotides that contain adenine and uridine the spectra can be easily interpreted. However, when cytosine is a component of the compounds being studied difficulties arise,[5] and until more studies are performed on low wavelength theory of cytosine $\pi \rightarrow \pi$ transitions, which in turn effect the $n \rightarrow \pi$ transitions around 240 nm, results from circular dichroic studies will be difficult to interpret. Circular dichroism has been suggested as a very powerful tool

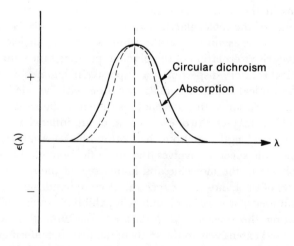

Figure 4–9. Schematic of the absorption spectrum and CD spectrum of a hypothetical substance.

in analysis of structure of nucleic acids.[5] Early studies were promising, showing differences between the conformation of RNA and DNA, in that RNA showed only a single positive band, whereas DNA showed two oppositely signed overlapping bands.[9] These results corresponded to studies with double-stranded RNA which showed two overlapping bands.[5] Corresponding studies with homopolymers of both double-stranded and single-stranded forms have not been extended, apart from one study,[113] because of the problems with low wavelength theory outlined previously. Circular dichroism is now considered by the majority of workers not to be as potentially powerful as first assumed.

Infrared Spectra

Infrared spectra measure the separation between vibrational energy levels of a molecule. Since the vibrational groups in a molecule show very definite wavelength characteristics, infrared spectroscopy has become one of the most useful tools for identification of compounds in organic chemistry. Infrared spectra have been used for investigation of the nucleic acid bases, nucleosides and nucleotides, but very few studies have been carried out on the polymeric molecules because of the complexities of the spectra and difficulty in interpretating the spectral bands. With long polymers the individual vibrations of the many similar groups present in the polymerized units of the polymer overlap, and a continuous absorption through most of the spectral region is seen instead of the well-defined peaks usually associated with an infrared spectrum. The continuous pattern shows peaks, but these are broad and irregular. Some studies have appeared which compare various types of nucleic acids[48,113] suggesting a limited use for infrared spectroscopy in macromolecular nucleic acid research.

Magnetic Resonance Spectra

Absorption of low frequency radiation in the radio frequency range of wavelength ($\lambda = 1$ cm and above) permits observation of three kinds of energy transition.

1. Pure rotational transitions between rotation levels of molecules in a gas phase. This is confined to gaseous substances and is not applicable to the majority of organic molecules or macromolecules.

2. Transitions corresponding to electron spin resonance resulting from the interaction of an external magnetic field with the spin of nuclei; this is called nuclear magnetic resonance.

3. Transitions between energy levels created by the interaction of an external magnetic field with the spin of unpaired electrons. This is called electron magnetic resonance, electron spin resonance or electron paramagnetic resonance. Electron spin resonance spectra are limited to molecules

with free radicals because of the unpaired electron phenomenon. Attachment of spin groups or production of free radicals by X-ray irradiation of molecules which normally have no free radicals extends the application of this method of spectroscopy.

Nuclear Magnetic Resonance Spectra

Nuclear magnetic resonance spectra yield detailed information on nucleic acids at the molecular level which cannot be obtained by other techniques used in investigations of the structure and interactions of the nucleic acids, such as ultracentrifugation, optical rotatory dispersion and optical density measurements. Nuclear magnetic resonance absorption occurs only with those nuclei which have non-zero spin, such as ^1H, ^2H, ^{13}C, ^{14}N, ^{17}O and ^{31}P. Since ^{12}C, ^{14}C, ^{16}O and ^{18}O have zero spin, nuclear magnetic resonance gives information about the H, N and P atoms in a molecule but not about the C or O atoms. The ^{13}C and ^{17}O atoms are so low in content in natural carbon and oxygen that they do not interfere. The relationship between the energy of interaction, E, the spin of the nucleus, I, and the magnetic field, H, is described by the equation

$$E = M\mu \left.\frac{H}{I}\right| \qquad (56)$$

where μ is the nuclear magnetic moment and M a quantum number. In the general consideration of spectroscopy we saw that $E = h\nu$. Assuming $\Delta M = 1$ and substituting in the equation, we have

$$\nu = \mu \frac{H}{hI} \qquad (57)$$

Experimentally the absorption is detected while the frequency, ν, of the incident radio frequency is kept constant and H is varied. Thus, absorption occurs when

$$H = \nu \frac{hI}{\mu} \qquad (58)$$

Nuclear magnetic resonance provides specific details of the molecular structure of molecules because of a number of readily observable effects. One of these is that the absorption peaks of nuclear magnetic resonance spectra in liquids are sharp compared with those in infrared and electronic spectra. The explanation for this difference is that the lifetime of a nuclear spin state is very long in comparison with electronic or vibrational energy states, so that interaction with nearby solvent molecules, which perturb the energy in a given state in a manner which varies from instant to instant, is averaged out to a constant value during the lifetime of the nuclear spin state. This situation does not exist for electronic or vibrational energy states. A second effect is that if nearby nuclei, which are joined by covalent bonds, also possess spin, interactions between the spins occur, splitting the individual absorption bands into several peaks. The number of peaks for any one nucleus is equal to the number of possible spin orientations of interacting nuclei. This provides a very sensitive measure of structure near absorbing nuclei. A third effect is the magnetic field to which the nucleus is subject. This is not the same as the applied field because of the shielding provided the nucleus by the surrounding electrons. However, the shielding depends on the chemical bonding of the nucleus being investigated. The protons of different groups—for example, $—NH_2$, $—CH_2—$, $—CH_3$ groups—will absorb at slightly different applied fields. The chemical shifts in the critical value of the magnetic field, H, for absorption may be used as a measure of the type of bonding of the absorbing atom to its neighbors.

Nuclear magnetic resonance data has been published on monomer units and various derivatives of the nucleic acids, and the structural analysis of the compounds has been worked out from the spectra. More recently, investigations have been extended to synthetic polynucleotides and intact nucleic acids. The advent of the high frequency 200 MHz machines has extended the range and possibilities of information available by nuclear magnetic resonance spectroscopy. Of particular interest to nucleic acid chemists have been investigations of monomer-monomer and monomer-polymer interactions. A recent review on this subject[95] describes in some detail information on 20 to 30 compounds. Fairly extensive nuclear magnetic resonance studies on the association of bases, nucleosides and nucleotides in solution have been carried out in a search for information concerning polymer association phenomena. Osmotic studies had shown previously that the association of bases and nucleosides in neutral aqueous solutions was an energetically favorable process and that interactions between nucleosides and purine bases were more favorable energetically than the corresponding interactions between pyrimidine bases and nucleosides.[117,118] These results have been confirmed and extended by nuclear magnetic resonance studies. Investigation of the association of purine nucleosides in aqueous solution showed that hydrogen bonding is not involved and that association occurs through a mechanism of vertical stacking.[10,102] Association occurs even when the potential sites for hydrogen bonding have been methylated. Investigation of adenosine and adenine ribonucleosides substituted at the N-6 position and comparison with 2' deoxyadenosine and 2'-substituted nucleosides suggested the existence of intra-molecular hydrogen bonding between the 2' hydroxyl group of the pentose and the N-3 atom of the purine ring. The existence of intra-molecular hydrogen bonding is supported by ultraviolet absorption and infrared spectroscopy data,[80,130] and it has been suggested that this may account for differences in conformation between polydeoxyribonucleotides and polyribonucleotides observed by optical rotatory dispersion spectra[119] (see Chapter 3). In contrast, base pairing of nucleosides and nucleoside derivatives has been shown to occur in non-aqueous solutions.[60] Investigation of these associations by nuclear magnetic resonance has provided supporting evidence for hydrogen bonding between specific bases,[90,103] giving direct support to the Watson-Crick base pairing model. Inoue and Aoyaga have studied the nuclear magnetic resonance spectra of three dinucleotides—ApGp, CpGp and UpGp—in D_2O solution.[56] Their spectra showed the H-6 proton signal of uracil in UpGp to be quite sharp, indicative of free rotation about the glycosidic linkage, whereas the rotation about the same linkage of the guanine base is strongly restricted, in contrast with ApGp in which a strong base-base interaction operates and rotation is completely restricted. CpGp was cited as an intermediate case. The hindered rotation of the 3' base residue in the unstacked dinucleotide UpGp indicated the importance of nucleotide sequence in stacking phenomena. These studies complemented previous ultraviolet and optical rotatory dispersion studies.

Since the original work of Inoue and Aoyaga similar studies on a number of other dinucleotides have appeared.[14,51] These few examples of monomer data indicate the supporting evidence nuclear magnetic resonance spectroscopy can provide to studies in other systems. Nuclear magnetic resonance spectroscopy of nucleic acids at the polymer level has been particularly useful. The formation of poly(A)·poly(U) and poly(A)·2 poly(U) co-polymers has been quantitatively followed from the intensity of nuclear magnetic resonance spectral lines.[76] Complexing and melting out of the co-polymers was shown to be completely reversible and the melting temperature dependent on the ionic concentration of the solvent. Investigation of poly(A) and poly(U) solutions containing various ratios of poly(A) and poly(U), examined as a function of temperature, showed that no appreciable complexing occurred above 65°C, but at 25°C poly(A)·poly(U) forms when the mole fraction of poly(U) is less than 0.5 and poly(A)·2 poly(U) forms when it is greater than 0.66, supporting the original ultraviolet absorbance data of Rich and Davies.[37,91] In studies on RNA and DNA a nuclear magnetic resonance spectral method for the determination of nearest neighbor base frequency ratios in DNA has been reported.[75] The method is based upon the thymine methyl group resonances in single-stranded DNA in neutral aqueous solution. From the spectra of dinucleotides containing thymine as the 5' neighbor and 3' neighbor distinct characteristics of the methyl group were observed. The methyl group gives one resonance when thymine is the 5' neighbor or when thymine has a 5' neighbor pyrimidine. However, when thymine has a 5' neighbor purine the methyl group resonances are shifted to higher field values at 20°C. This effect decreases with higher temperatures. The temperature shift supports the observation that purines and pyrimidines exhibit stacking interactions at low temperatures in neutral aqueous solutions. From these observations the shifts of the methyl group resonances of dApT and dGpT have been ascribed to non-random stacking of the purine and pyrimdine bases, with the methyl group lying within the ring field current of the purine as the favored configuration. Nuclear magnetic resonance studies have been used as a probe of transfer RNA structure,[19,106] providing data useful in model building (see Chapter 2). Even from this brief account it is obvious that nuclear magnetic resonance spectra provide unique opportunities for investigation of structural phenomena of the nucleic acids.

Electron Spin Resonance

Electron spin resonance spectra arise from the flip-flop of the electron spin an an external magnetic field when an electromagnetic wave of the correct frequency hits a molecule with unpaired electrons. The electron spin resonance technique is used to determine not only the number of unpaired electrons in a sample but also the internal magnetic fields which affect the energy levels. In nucleic acids, since the naturally occurring molecules do not have any free radical states, as far as is known, electron spin resonance data has been examined on X-irradiated DNA,[20,89] which results in brief free radical states. However, the more recent studies on nucleic acids have utilized the technique of attachment of a spin label to a selected portion of the nucleic acid molecule to be studied and the mobility of that portion observed by measurement of the relaxation or tumbling of the attached spin label.[44] Hoffman, Schofield and Rich have used this technique to study the denaturation of transfer RNA.[49] They attached a nitroxide spin label to the α-amino group of valyl transfer RNA and measured the spin label as a function of temperature and ionic strength. They showed that there was an abrupt transition in the nature of the spin motion at the same temperature as the midpoint of the optical density melting curve and that the transition was sensitive to ionic strength. When denaturing agents such as dimethylsulfoxide or urea were added to the solution, the temperature of transition of the spin motion was lower than the change in the melting temperature as measured by the optical density melting curve method, suggesting that the part of the molecule to which the spin label was attached (the aminoacyl end) denatured before the rest of the molecule.

Since electron spin resonance is also a relatively new technique and the apparatus is not generally available, there has been relatively little study of the nucleic acids by this method. However, it has already proved to be a very useful probe of nucleic acid structure.

The methods and results obtained by the physicochemical approach to nucleic acid structure are not comprehensively treated in this monograph and could not be so in such a brief review. Deliberately not included were certain methods such as flow dichroism,[12] fluorescence polarization,[81] and polarographic measurements of nucleic acids,[86] all of which have been investigated in recent years. Hopefully this introduction has provided some insight into the chemistry and physics of DNA and RNA and indicated some future possibilities for research.

REFERENCES

1. Archibald, W. J., J. Phys. Colloid. Chem. *51*, 1204 (1947).
2. Aten, J. B. T., and Cohen, J. A., J. Mol. Biol. *12*, 537 (1965).
3. Bancroft, F. C., and Freifelder, D., J. Mol. Biol. *54*, 537 (1970).
4. Beaven, G. H., Holiday, E. R., and Johnson, E. A., in *The Nucleic Acids* (E. Chargaff and J. N. Davidson, Eds.) Academic Press, New York, Vol. 1, p. 493 (1955).
5. Beychok, S., Science *154*, 1288 (1966).
6. Bloomfield, V., and Zimm, B. H., J. Chem. Phys. *44*, 315 (1966).
7. Boedtker, H., and Simmons, N. S., J. Amer. Chem. Soc. *80*, 2550 (1958).
8. Brahms, J., Maurizot, J. C., and Michelson, A. M., J. Mol. Biol. *25*, 465 (1967).
9. Brahms, J., and Mommaerts, W. F. H. M., J. Mol. Biol. *10*, 73 (1964).
10. Broom, A. D., Schweitzer, M. P., and Ts'o, P. O. P., J. Amer. Chem. Soc. *89*, 3612 (1967).
11. Burgi, E., and Hershey, A. D., J. Mol. Biol. *3*, 458 (1961).
12. Callis, P. R., and Davidson, N., Biopolymers *7*, 335 (1969).
13. Cannon, M. R., and Fenske, M. R., Ind. Eng. Chem., Anal. Ed. *10*, 297 (1938).
14. Chan, S. I., and Nelson, J. H., J. Amer. Chem. Soc. *91*, 168 (1969).
15. Cheng, P.-Y., Biochem. Biophys. Res. Commun. *33*, 746 (1968).
16. Cheng, P.-Y., and Schachman, H. K., J. Polymer Sci. *16*, 19 (1955).
17. Clark, N. C., Lunacek, J. H., and Benedek, G. B., Amer. J. Phys. *38*, 575 (1970).
18. Cohen, G., and Eisenberg, H., Biopolymers *6*, 1077 (1968).
19. Cohn, M., Danchin, A., and Grunberg-Manago, M., J. Mol. Biol. *39*, 199 (1969).
20. Cook, J. B., and Wyard, S. J., Nature *210*, 526 (1966).
21. Couette, M., Ann. Chim. Phys. *21*, 433 (1890).
22. Crothers, D. M., Biopolymers *6*, 1391 (1968).
23. Crothers, D. M., Kallenbach, N. R., and Zimm, B. H., J. Mol. Biol. *11*, 802 (1965).
24. Crothers, D. M., and Zimm, B. H., J. Mol. Biol. *12*, 525 (1965).
25. Davies, D. R., Ann. Rev. Biochem. *36*, 321 (1967).
26. Davis, R. C., and Tinoco, I., Jr., Biopolymers *6*, 223 (1968).
27. Davis, R. W., and Davidson, N., Proc. Nat. Acad. Sci. (Wash.) *60*, 243 (1968).
28. de Duve, C., Berthet, J., and Beaufay, H., Prog. Biophys. Biophys. Chem. *9*, 326 (1959).
29. Donnan, F. G., Z. Elektrochem. *17*, 572 (1911).
30. Dubin, S. B., Benedek, G. B., Bancroft, F. C., and Freifelder, D., J. Mol. Biol. *54*, 547 (1970).
31. Dubin, S. B., Lunacek, J. H., and Benedek, G. B., Proc. Nat. Acad. Sci. (Wash.) *57*, 1164 (1967).
32. Eigner, J., and Doty, P., J. Mol. Biol. *12*, 549 (1965).
33. Einstein, A., Ann. Physik. *19*, 289 (1906); *34*, 591 (1911).
34. Eisenberg, H., J. Polymer Sci. *25*, 257 (1957).
35. Eizner, Y. E., and Ptitsyn, O. B., Vysokomolek. Soedineniya *4*, 1725 (1962).
36. Elson, E. L., Scheffler, I. E., and Baldwin, R. L., J. Mol. Biol. *54*, 401 (1970).
37. Felsenfeld, G., Davies, D. R., and Rich, A., J. Amer. Chem. Soc. *79*, 2023 (1957).
38. Freifelder, D., J. Mol. Biol. *54*, 567 (1970).
39. Gillespie, D., Methods in Enzymology *12B*, 641 (1968).
40. Gillespie, D., and Spiegelman, S., J. Mol. Biol. *12*, 829 (1965).
41. Goel, N. S., and Maitra, S. C., J. Theor. Biol. *23*, 87 (1969).
42. Gruenwedel, D. W., and Hsu, C.-H., Biopolymers *7*, 557 (1969).
43. Hall, B. D., and Spiegelman, S., Proc. Nat. Acad. Sci. (Wash.) *47*, 137 (1961).
44. Hamilton, C. L., and McConnell, H. M., in *Structural Chemistry and Biology* (A. Rich and N. Davidson, Eds.), W. H. Freeman and Co., San Francisco, p. 115 (1968).
45. Hearst, J. E., and Stockmayer, W. H., J. Chem. Phys. *37*, 1425 (1962).
46. Hermans, J., Jr., and Hermans, J. J., Proc. Koninkl. Ned. Akad. Wetenschap. B*61*, 324 (1958).
47. Herzog, R. O., Illig, R., and Kudar, H., Z. Phys. Chem. A*167*, 329 (1934).
48. Higuchi, S., Tsuboi, M., and Iitaki, Y., Biopolymers *7*, 909 (1969).
49. Hoffman, B. M., Schofield, P., and Rich, A., Proc. Nat. Acad. Sci. (Wash.) *62*, 1195 (1969).

50. Houwink, R., J. Prakt. Chem. *157*, 15 (1941).
51. Hruska, F. E., and Danyluk, S. S., Biochim. Biophys. Acta *157*, 238 (1968).
52. Huggins, M. L., J. Amer. Chem. Soc. *64*, 2716 (1942).
53. Inman, R. B., J. Mol. Biol. *18*, 464 (1966).
54. Inman, R. B., J. Mol. Biol. *25*, 209 (1967).
55. Inman, R. B., J. Mol. Biol. *28*, 103 (1967).
56. Inoue, Y., and Aoyagi, S., Biochem. Biophys. Res. Commun. *28*, 973 (1967).
57. Inoue, Y., Aoyagi, S., and Nakanishi, K., J. Amer. Chem. Soc. *89*, 5701 (1967).
58. Inoue, Y., Masuda, M., and Aoyagi, S., Biochem. Biophys. Res. Commun. *31*, 577 (1968).
59. Inoue, Y., and Satoh, K., Biochem. J., *113*, 843 (1969).
60. Katz, L., and Penman, S., J. Mol. Biol. *15*, 220 (1966).
61. Kleinschmidt, A. K., and Zahn, R. K., Z. Naturforsch. *146*, 770 (1959).
62. Kraemer, E. O., Ind. Eng. Chem. *30*, 1200 (1938).
63. Kraut, J., Ann. Rev. Biochem. *34*, 247 (1965).
64. Lang, D., J. Mol. Biol. *54*, 557 (1970).
65. Lang, D., Bujard, H., Wolff, B., and Russell, D., J. Mol. Biol. *23*, 163 (1967).
66. MacGillivray, A. D., and McMullen, A. I., J. Theor. Biol. *12*, 260 (1966).
67. Mandelkern, L., and Flory, P. J., J. Chem. Phys. *20*, 212 (1952).
68. Mark, H., in *Der Feste Körper*, Hirzl, Leipzig, p. 103 (1938).
69. Marmur, J., and Doty, P., Nature *183*, 1427 (1959).
70. Marmur, J., and Doty, P., J. Mol. Biol. *3*, 585 (1961).
71. Marmur, J., and Doty, P., J. Mol. Biol. *5*, 109 (1962).
72. Marmur, J., Rownd, R., and Schildkraut, C. L., Prog. Nucleic Acid Res. Mol. Biol. *1*, 231 (1963).
73. Martin, R. G., and Ames, B. N., J. Biol. Chem. *236*, 1372 (1961).
74. Massie, H. R., and Zimm, B. H., Biopolymers *7*, 475 (1969).
75. McDonald, C. C., Phillips, W. D., and Azar, J. L., J. Amer. Chem. Soc. *89*, 4166 (1967).
76. McDonald, C. C., Phillips, W. D., and Penman, S., Science *144*, 1234 (1964).
77. Mehl, J. W., Oncley, J. L., and Simha, R., Science *92*, 132 (1940).
78. Meselson, M., and Stahl, F. W., Proc. Nat. Acad. Sci. (Wash.) *44*, 671 (1958).
79. Meselson, M., Stahl, F. W., and Vinograd, J., Proc. Nat. Acad. Sci. (Wash.) *43*, 581 (1957).
80. Michelson, A. M., Ann. Rev. Biochem. *30*, 133 (1963).
81. Millar, D. B., and MacKenzie, M., Biochem. Biophys. Res. Commun. *23*, 724 (1966).
82. Miller, O. L., Jr., Beatty, B. R., Hamkalo, B. A., and Thomas, C. A., Jr., Cold Spring Harb. Symp. Quant. Biol. *35*, 505 (1970).
83. O'Konski, C. T., and Haltner, A. J., J. Amer. Chem. Soc. *78*, 3604 (1956).
84. Opschoor, A., Pouwels, P. H., Knijnenburg, C. M., and Aten, J. B. T., J. Mol. Biol. *37*, 13 (1968).
85. Osborne, C. F., J. Chem. Phys. *44*, 2735 (1966).
86. Palečٍek, E., J. Mol. Biol. *20*, 263 (1966).
87. Perrin, F., J. Phys. Radium *5*, 497 (1934).
88. Perrin, F., J. Phys. Radium *7*, 1 (1936).
89. Pershan, P. S., Shulman, R. G., Wyluda, B. J., and Eisinger, J., Science *148*, 378 (1965).
90. Pople, J. A., Schneider, W. G., and Bernstein H. J., in *High Resolution Nuclear Magnetic Resonance*, McGraw-Hill, New York (1959).
91. Rich, A., and Davies, D. R., J. Amer. Chem. Soc. *78*, 3548 (1956).
92. Rolfe, R., and Meselson, M., Proc. Nat. Acad. Sci. (Wash.) *45*, 1039 (1959).
93. Rosenberg, A. H., and Studier, F. W., Biopolymers *7*, 765 (1969).
94. Ross, P. D., and Scruggs, R. L., Biopolymers *6*, 1005 (1968).
95. Rowe, J. J. M., Hinton, J., and Rowe, K. L., Chem. Rev. *70*, 27 (1970).
96. Sarkar, P. K., and Yang, J. T., J. Biol. Chem. *240*, 2088 (1965).
97. Schachman, H. K., *Ultracentrifugation in Biochemistry*, Academic Press, New York (1959).
98. Scheffler, I. E., Elson, E. L., and Baldwin, R. L., J. Mol. Biol. *36*, 291 (1968).
99. Scheffler, I. E., and Sturtevant, J. M., J. Mol. Biol. *42*, 577 (1969).
100. Scheraga, H. A., and Mandelkern, L., J. Amer. Chem. Soc. *75*, 179 (1953).
101. Schildkraut, C. L., Marmur, J., and Doty, P., J. Mol. Biol. *4*, 430 (1962).
102. Schweitzer, M. P., Chan. S. I., and Ts'o, P. O. P., J. Amer. Chem. Soc. *87*, 5241 (1965).

103. Shoup, R. R., Miles, H. T., and Becker, E. D., Biochem. Biophys. Res. Commun. 23, 194 (1966).
104. Simha, R., J. Phys. Chem. 44, 25 (1940).
105. Sinsheimer, R. L., Scott, J. F., and Loofbourow, J. R., J. Biol. Chem. 187, 313 (1950).
106. Smith, I. C. P., Yamane, T., and Shulman, R. G., Science 159, 1360 (1968).
107. Staudinger, H., and Heuer, W., Chem. Ber. 63, 222 (1930).
108. Staudinger, H., and Nodzu, R., Chem. Ber. 63, 721 (1930).
109. Stokes, G., Sir, Trans. Cambridge Phil. Soc. 8, 287 (1847), 9, 8 (1851).
110. Sueoka, N., Proc. Nat. Acad. Sci. (Wash.) 45, 1480 (1959).
111. Svedberg, T., and Pedersen, K. O., in *The Ultracentrifuge*, Oxford University Press, London (1940), Johnson Reprint Corp., New York.
112. Szybalski, W., in Methods in Enzymology 12B, 330 (1968).
113. Thomas, G. J., Jr., Biopolymers 7, 325 (1969).
114. Thrower, K. J., and Peacocke, A. R., Biochem. J., 109, 543 (1968).
115. Tinoco, I., Jr., J. Amer. Chem. Soc. 82, 4785 (1960).
116. Toal, J. N., Rushinsky, G. W., and Pratt, A. W., Anal. Biochem. 23, 60 (1968).
117. Ts'o, P. O. P., and Chan, S. I., J. Amer. Chem. Soc. 86, 4176 (1964).
118. Ts'o, P. O. P., Melvin, I. S., and Olson, A. C., J. Amer. Chem. Soc. 85, 1289 (1963).
119. Ts'o, P. O. P., Rappaport, S. A., and Bollum, F. J., Biochemistry 5, 4153 (1966).
120. Tuan, D. Y. H., and Bonner, J., J. Mol. Biol. 45, 59 (1969).
121. Ubbelohde, L., Ind. Eng. Chem., Anal. Ed. 9, 85 (1937).
122. Vinograd, J., and Bruner, R., Biopolymers 4, 131 (1966).
123. Vinograd, J., and Bruner, R., Biopolymers 4, 157 (1966).
124. Vinograd, J., Bruner, R., Kent, R., and Weigle, J., Proc. Nat. Acad. Sci. (Wash.) 49, 902 (1963).
125. Voet, D., and Rich, A., Prog. Nucleic Acid Res. Mol. Biol. 10, 183 (1970).
126. Vournakis, J. N., Scheraga, H. A., Rushizky, G. W., and Sober, H. A., Biopolymers 5, 33 (1966).
127. Warshaw, M. M., and Tinoco, I., Jr., J. Mol. Biol. 20, 29 (1966).
128. Westmoreland, B. C., Szybalski, W., and Ris, H., Science 163, 1343 (1969).
129. Wetmur, J. G., and Davidson, N., J. Mol. Biol. 31, 349 (1968).
130. Witzel, H., Ann. Chem. 635, 182 (1960).
131. Wolfe, F. H., Oikawa, K., and Kay, C. M., Can. J. Biochem. 47, 977 (1969).
132. Zimm, B. H., in *Fractions*, Beckman Instruments, Palo Alto, California, No. 3 (1965).
133. Zimm, B. H., and Crothers, D. M., Proc. Nat. Acad. Sci. (Wash.) 48, 905 (1962).

INDEX

Acyl hydrazides of cytosine, 40
Animal and plant DNA, 34
1-β-D-Arabinofuranosylthymine, 10
1-β-D-Arabinofuranosyluracil, 10
Arabinose, 9
Archibald method of sedimentation analysis, 113-114

Bacterial DNA, 34
  conformation, 34
  episomes, 34
  molecular weight, 34
  plasmids, 34
  satellite DNA, 34
Band sedimentation, 115
  determination of sedimentation and diffusion coefficients by, 115
Base composition and buoyant density relationship, 114
Base stacking, 57, 92, 134
Birefringence, 122
Borohydride, reduction of oxidized nucleosides by, 14

Capillary viscometers, 103-105
  Ostwald viscometer, 104
  Poiseuille's equation, 104
Chemical synthesis of deoxyoligonucleotides, 73-85
Chemical synthesis of polynucleotides, 73-88
Chloroplast DNA, 33
Chromosome organization. See Nuclear DNA.
Circular dichroism and polynucleotide conformation, 90
Concatenated DNA's, 60

Condensation reactions, 74-79
Configuration of cytidine, 42
  DNA double helix and, 56
  X-ray diffraction study, 42
Configuration of the nucleic acids, 42-62
  DNA, 55-62
  messenger RNA, 47-48
  ribonucleic acids, 42-55
  ribosomal RNA, 42-47
  5s RNA, 52-55
  transfer RNA, 48-52
Couette viscometer, 105
Cytokinin(s), 19-26
Cytoplasmic DNA, 33-34
  chloroplast, 33
  conformation, 33
  mitochondrial, 33
  organization in the mitochondrion, 33

Density gradient centrifugation, 114
  determination of s, 116
  preformed gradients, 115-116
  relationship to base composition, 114
Deoxyoligonucleotides, chemical synthesis of, 73-85
Deoxyribonucleic acid, 29-35
  and protein, 60-61
    acidic proteins, 61
    histones, 61
    protamines, 61
  bacterial, 34
  base components of, 10-12
  base pairing, 56-57
  concatenated DNA's, 60
  double-stranded helical model, 55-59
  ethidium bromide, intercalation of, 60
  four-stranded models, 58
  occurrence, 29
  organization in the chromosome, 59-61

Deoxyribonucleic acid *(Continued)*
  permuted and non-permuted DNA's, 60
  protein linkers, 59
  reiterated sequences, 60
  repetitive sequences. See Satellite DNA.
  single-stranded circular bacteriophage DNA's, 59, 60
  stacking forces, 57
DNA polymerase and multiplication of oligodeoxynucleotides, 85
  synthesis of homo and copolymers, 73
DNA sequence determination, 35, 40-41
  chemical degradation, 40
  cystosine acyl hydrazides, 40
  electron microscopy and base modification, 40
  modification of DNA by
    2.4-dinitrophenylhydrazine deamination, 41
    formaldehyde, 41
    formamide, 41
    hydrazinolysis, 41
    hydrogen peroxide, 41
    hydroxylamine, 41
    osmium tetroxide, thymine reaction with, 40, 41
    partial acid hydrolysis, 40
    permanganate oxidation of pyrimidines, 41
    semicarbazide, cytidine reaction with, 41
    purine catalogue (clusters), 41
    pyrimidine catalogue (clusters), 40, 41
DNA terminal transferase, 73
DNases specific, 41
  E. coli exonuclease III, 41
2-deoxy-D-ribose, 6
Dicyclohexylcarbodi-imide. See Phosphorylating agents.
Diffusion coefficient, measurement of, 112
Donnan effect. See Osmotic pressure.

Einstein's equation (viscosity), 106
Electron microscopy, 40
Electron microscopy and molecular weight determination, 120-121
Electron spin resonance, 135
  spin labels, use of in nucleic acids, 135
  tRNA structure and, 135
Episomes, 34
Ethidium bromide, intercalation of, 60

Falling sphere viscometers, 106
Flow birefringence, 122-124
  apparatus, description of, 123-124
  measurement, description of, 123-124

Frictional coefficient, f, 103, 111, 118

Gene synthesis of alanine tRNA, 85-88
Glucose, 9
  in bacteriophage DNA, 11
N-glycoside bonds, 6

Histones, 61
Homochromatography, 37
Hydrazinolysis, 41
Hydrodynamic properties, 101-124
  flow birefringence, 122-124
  sedimentation analysis, 108-117
  sedimentation and viscosity relationship, 117-120
  viscosity, 101-108
Hydrogen bonding, 91, 128, 134
5-hydroxy-1-α-(6-0-β-D-glucopyranosyl-D-glucose) methylcytosine, 11
5-hydroxymethylcytosine, 10, 11
  glucosylation of, 11
5-hydroxymethyluracil, 11
Hyperchromicity, 127

Informosomes, 25
Infrared spectra, 131
  comparative studies of nucleic acids, 131
  of bases, 131
Inosine, cyanoethylation of, 26
Intrinsic viscosity, 107
Ionophoresis, 137
N^6-(Δ^2isopentenyl) adenosine, 13
  chemical synthesis of, 18
  identification of, 15-20

Joining reactions with oligodeoxynucleotides, 85

Kleinschmidt technique, 120

λ bacteriophage, sequence, 35
Laminar flow, 103

Magnetic Resonance Spectra, 131-135
  electron spin resonance, 135
  evidence for hydrogen bonding, 134
  general principles, 132-133

Magnetic Resonance Spectra (*Continued*)
  monomer-monomer interactions, 133
  monomer-polymer interactions, 133
  Nuclear Magnetic Resonance, 132-135
Mandelkern-Scheraga-Flory constant, β, 118-119
Mesitylene sulfonylchloride. See phosphorylating agents.
Messenger RNA and base composition, 25
  configuration, 47-48
  informosomes, 48
  molecular weight, 25, 48
  occurrence, 25
5-Methylcytosine, 10
N^6-methyladenine, 12
N^2-methylguanine, 12
2'-O-Methyl-1-β-D-ribofurnanosyladenine, 9
5-Methyluracil. See Thymine.
Minor RNA species, 28
  7s bacterial RNA, 28
  7-10s mammalian species, 28
Mitochondria, 33
  DNA, 33
  messenger RNA, 33
  ribosomes, 33
  transfer RNA, 33
Model polynucleotides, studies with synthetic, 88-95
  monomer-polymer interactions, 91-93
  oligonucleotide-polymer interactions, 91-93
  polymer-polymer interactions, 88-91
  single strand polymer interaction, 93-95
Molecular weight and viscosity, 108
Molecular weights by electron microscopy, 120-121
  density per unit length, 121
  structural correlations, 121
Monomer-polymer interaction,
  hydrogen bond and, 91
  stacking of bases, 92
  three strand/two strand helices, 91-92
MS2 bacteriophage, 28

Nuclear DNA, 31, 32
  base composition, 32
  chromosome organization, 31, 59-61
  molecular weight, 31
Nuclear Magnetic Resonance Spectra,
  base stacking and sequence, 134
  nucleoside identification by, 14, 15, 19
  poly (A). poly (U) complexes, and, 134
  polynucleotide conformation, 90
Nucleic acids, brief history of, 1-4
  molecular structure of, 22-62
    structural components of, 6-19
      bases, 10-19
      sugars, 6-10

Oligodeoxynucleotides, multiplication of, 85
  DNA polymerase and, 85
Oligonucleotide-polymer interaction, 92-93
  double strand and triple strand complexes, 92-93
Oncogenic viruses, 28
Optical rotatory dispersion and circular dichroism, 129-131
  as probes of molecular structure, 90, 129
  oligonucleotides, and, 129-130
  RNA and DNA, and, 131
Osmotic pressure, 99-101
  and molecular weight, 100
  and nucleic acids, 100
  Donnan effect, 101
Ostwald viscometer, 104

Pancreatic RNase and sequencing, 35, 36, 37
Periodate, oxidation of nucleosides, 14, 18
Permuted and non-permuted DNA's, 60
ØX174 bacteriophage, sequence, 35
Phosphorylating agents,
  dicyclohexylcarbodi-imide, 74
  mesitylene sulfonylchloride, 80
  p-toluenesulfonyl chloride, 74
Plant DNA. See Animal and plant DNA.
Plasmids, 34
Poiseuille's equation, 104
Polynucleotides, chemical synthesis of, 73-88
  condensation reactions, 74-79
  stepwise synthesis, 79-82
Poly [d(A-T)] alternating copolymer, chemical synthesis of, 75-76
poly (A), double-stranded acid structure of, 93
poly (C), double-stranded acid structure of, 94
poly (G), aggregation of, 94
poly (I) three-strand structure of, 94
poly (2'-0-methyladenylic acid), 90
poly (2'-0-methylcytidylic acid), 89-90
poly (5-methyldeoxycytidylic acid), 90
Polydeoxynucleotides and the genetic code, 82
Polymer interactions, single-stranded,
  poly (A), double-stranded acid structure, 93
  poly (C), double-stranded acid structure, 94
  poly (G), aggregation, 94
  poly (I), three-strand structure, 94
Polymer-polymer interactions, double helix formation, 88-89
  helix stability and bases, 90

Polymer-polymer interactions (*Continued*)
  helix stability and hydrogen bonds, 91
  helix stabilization and sugars, 89
Polynucleotide conformation,
  circular dichroism and, 90
  Nuclear Magnetic Resonance and, 90
Polynucleotide ligase. See T4 polynucleotide ligase.
Polynucleotide phosphorylase, 69-72
  arsenolysis, 72
  nucleoside diphosphates as substrates (specificity), 71
  occurrence, 70
  phosphorolysis, 72
  polymerization reaction, 70, 71
  preparation and assay, 70
  primer requirement, 70
  transnucleotidation, 72
Polyribonucleotides, chemical synthesis of, 73
Protamines, 61
Protecting groups, 74-75
  acetyl, 74
  β-cyanoethyl, 75
  N-anisoyl, 80
  N-benzoyl, 74, 75
  p-methoxytrityl, 75
Pseudouridine, 13
  cyanoethylation of, 26
  identification of, 14, 15

Qβ bacteriophage, 28

Relative viscosity, 107
Reovirus, 29
5-β-D-Ribofuranosyluracil. See Pseudouridine.
Ribonuclease, synthesis of oligonucleotides, 73
  See also pancreatic ribonuclease and T1 ribonuclease.
Ribonucleic acid, alkaline hydrolysis of, 6
  branching of, 6
  composition of, 22-29
  occurrence of, 22-29
  primary structure of, 22-29
RNA of complementary sequence, synthesis of, 85
RNA polymerase and, 85
RNA polymerase, polyribonucleotide synthesis, 73
RNA sequence determination, chemical modification of bases, 36
  general method, 35-36
  pancreatic RNase, 35
  sequence of tRNA's, 35-36
  T1 RNase (Takadiastase), 35

RNA viruses, bacterial, 28
  base composition, 28
  double-stranded RNA, 29
  plant and animal, 29
  sequence, 28
5s RNA, 27-28
  base composition, 27
  base paired structures of (possible), 39
  base pairing content and, 53
  binding sites to 50s subunit, 55
  occurrence, 27
5s RNA sequencing, 29, 36-40
  carbodi-imide derivative of uracil and guanine, 36, 38
  complete sequence, 39
  methylation of guanylic acid, 36, 38
  modification of bases and enzyme hydrolysis, 36
  pancreatic and T1 RNase digests, 36, 37
D-Ribose, 6
  2'-0 methyl derivative of, 9
Ribosomal RNA, 23-24
  base composition of, 24
  formaldehyde, effect on, 43, 44
  molecular weight of, 9
  nucleotide sequence of, 24
  occurrence of, 23
Ribosome, models of, 43-47
  classification of, 23
  electron microscope studies of, 43-44
  helix content of, 43
  messenger RNA and, 44
  physical properties, 43, 44
  ribonuclease, action on, 43
  5s RNA and, 45
  subunits of, 23
  transfer RNA and, 44
Rotary diffusion, 122-123
  coefficient of, 123
Rotating cylinder viscometers, 105-106
R17 bacteriophage, 28

Sanger's sequencing technique, 37
  homochromatography, 37
  ionophoresis, cellulose acetate, 37
Satellite DNA, 32-33, 34
  isolation, 32, 33
  occurrence, 32
  repetitive sequences, 32, 33
Schlieren optics, 117
Sedimentation and viscosity, relationship between, 117-120
  for circular double stranded DNA, 120
  for linear double stranded DNA, 117-119
Sedimentation analysis, 108-109
  general principles, 108-109
Sedimentation coefficient, s, 111

Sedimentation equilibrium method, 109-111
  Boltzmann distribution, 110
  molecular weight, determination by, 110-111
Sedimentation velocity method, 111-113
  general principles, 111
  molecular weight, determination by, 112
  Svedberg equation, 112
Sequence studies, 35-41
  DNA's, 40-41
  RNA's, 35-40
Shear stress, 105
Single-stranded DNA, 59, 60
Specific viscosity, 107
Spectroscopic properties, 124-135
  general theory, 124-126
Spleen phosphodiesterase, 6
Spongosides, 9
  spongothymidine, 10
  spongouridine, 10
S13 bacteriophage sequence, 35
Stacking forces, 57, 128-129
Stepwise chemical synthesis, 79-82
Stokes' law, 103
Sucrose gradients, 115
Svedberg equation, 112
  Svedberg unit S, 112
Synthesis on a polymeric support, 82-85
  condensation reactions, 84
  linkage of first nucleoside, 84
  preparation of support, 83
Synthetic model polynucleotides, 69-95
  studies with, 88-95
  synthesis of, 69-88

Takadiastase. See T1 ribonuclease.
Thermal denaturation, 128-129
  application to structural studies, 128
  hydrogen bonding, and, 128
  methods of measurement, 128
  relation to base composition, 128
  stacking forces, and, 128-129
Thio bases, 19
Thymine, 10
T4 polynucleotide ligase, 88
T7 bacteriophage, sequence, 35
T1 ribonuclease and sequencing, 35, 36, 37

Tobacco mosaic virus, 29
p-Toluenesulfonyl chloride. See Phosphorylating agents.
Transfer RNA,
  carbodi-imide reaction and, 51
  clover leaf model, 48, 49
  cyanoethylation and, 51
  hydrogen exchange, and, 49
  Levitt model, 51-52
  melting and, 50-51
  minor nucleoside components of, 12, 19
  molecular weight of, 9, 26
  monoperphthalic acid (adenosine oxidation) and, 50
  occurrence, 25
  sequence, 26, 27
Trinucleotide, repeating, chemical synthesis of, 80

Ultracentrifuge, analytical, 116-117
  cell, description of, 116
  Schlieren optics, 117
Ultraviolet and visible spectra, 126-129
  applications to nucleic acids, 126-127

Viral DNA, 34-35
  bacteriophages, 35
  occurrence, 34
  sequence studies, 35
Viral RNA, 28, 29
  occurrence, 28
Viscosity, 101-108
  definition, Newton, 101
  general considerations, 101, 102
  laminar flow, 103
  molecular weight and, 108
  Newtonian viscosity, 103
  sedimentation and, 117-120
  Stokes' law, 103
Viscosity of macromolecular solutions, 106-108

Worm-like chain model, 118

THE LIBRARY
UNIVERSITY OF CALIFORNIA
San Francisco

FEB 11 1970